Technical
English
4
Workbook

Christopher Jacques

Pearson Education Limited

Edinburgh Gate

Harlow

Essex CM20 2JE

England and Associated Companies throughout the world.

www.pearsonelt.com

© Pearson Education Limited 2011

The right of Christopher Jacques to be identified as author of this Work has been asserted by him in accordance with the Copyright, Designs and Patents Act 1988.

First published 2011

Eighth impression 2019

ISBN: 978-1-4082-6800-1 (with key pack)
ISBN: 978-1-4082-6801-8 (without key pack)

Set in Cheltenham

Printed in Great Britain by Ashford Colour Press Ltd.

Acknowledgements

We would like to dedicate this book to the memory of David Riley, whose tireless professionalism contributed so much to its creation and success.

The publishers and author would like to thank the following for their invaluable feedback, comments and suggestions, all of which played an important part in the development of the course: Eleanor Kenny (College of the North Atlantic, Qatar), Julian Collinson, Daniel Zeytoun, Millie and Terry Sutcliffe (all from Higher Colleges of Technology, UAE), Dr Saleh Al-Busaidi (Sultan Qaboos University, Oman), Francis McNeice, (IFOROP, France), Michaela Müller (Germany), Malgorzata Ossowska-Neumann (Gdynia Maritime University, Poland), Gordon Kite (British Council, Italy), Wolfgang Ridder (VHS der Stadt Bielefeld, Germany), Stella Jehanno (Sentre d'Étude des Langues / Centre de Formation Supérieure d'Apprentis, Chambre de Commerce et d'Industrie de l'Indre, France) and Nick Jones (Germany).

We are grateful to the following for permission to reproduce copyright material:

Extract 4.1 adapted from 'About Diamond, What is Synchrotron Light used for?', http://www.diamond.ac.uk. Reproduced by kind permission of Diamond Light Source Ltd, copyright © 2011.

Illustration acknowledgements

HL Studios

Photo acknowledgements

The publisher would like to thank the following for their kind permission to reproduce their photographs:

(Key: b-bottom; c-centre; l-left; r-right; t-top)

4 Art Directors and TRIP Photo Library: Helene Rogers (l). **Science Photo Library Ltd:** Scott Camazine (r). **5 Tony Waltham/Geophotos. 8 Art Directors and TRIP Photo Library:** Helene Rogers. **12 Photolibrary. com:** Radius Images. **14 Alamy Images:** Doug Steley A. **22 Alamy Images:** Jan Nevill. **26 Rex Features:** Dmitry Beliakov. **28 Rex Features:** Mark Milstein. **30 Alamy Images:** Eye Ubiquitous. **32 Robert Harding World Imagery:** Tony Waltham. **35 Alamy Images:** Frances M. Robert. **38 Photolibrary.com:** Ian Lishman / Juice Images. **40 Science Photo Library Ltd:** Massimo Brega. **43 Photolibrary.com:** Jian Chen. **46 Alamy Images:** Jan Stromme. **52 Alamy Images:** Ace Stock Limited. **54 Press Association Images:** AP Photo / Abel Szalontai. **56 Getty Images:** AFP. **60 University of Manchester:** (c)

Cover images: *Front:* **Alamy Images:** Gary Moseley

All other images © Pearson Education

Every effort has been made to trace the copyright holders and we apologise in advance for any unintentional omissions. We would be pleased to insert the appropriate acknowledgement in any subsequent edition of this publication.

Contents

Unit 1 Innovations

4 **1.1 Eureka!**
Inventions
Past continuous and past simple
Present perfect continuous + for/since

5 **1.2 Smart drilling**
Article on crude oil
Past participial phrase at the beginning of
a sentence

6 **1.3 Lasers**
How an X-ray machine works
Signals in a lecture: change of topic, refer to
a visual, change of speaker

Unit 2 Design

8 **2.1 Spin-offs**
Description of barcodes
Present simple and past simple passive
Function: to + infinitive; for + verb + -ing;
that/which

9 **2.2 Specifications**
Design specification of a mobility scooter
Present participial phrases

10 **2.3 Properties**
Brainstorming meeting about a building
renovation
Technical presentation
Signals in a meeting: inviting suggestions,
introducing ideas, praising contributions

12 **Review Unit A**

Unit 3 Systems

14 **3.1 Product recall (1)**
Narrative of a pollution incident
Present continuous passive
Prepositional/Adverbial phrases to express
unlikelihood

15 **3.2 Product recall (2)**
News report of a car recall
Linking phrases using relative pronouns and
present participial phrases

16 **3.3 Controls**
Description of autopilot and windvane control
systems
Vocabulary of control systems and their
operation

Unit 4 Procedures

18 **4.1 Shutdown**
Description of a technological installation
Phrases to describe purpose and cause

19 **4.2 Overhaul**
Narrating an industrial accident
Phrasal verbs

20 **4.3 Demonstration**
Description of two welding processes
while/as + verb + -ing / present continuous
for simultaneous actions

22 **Review Unit B**

Unit 5 Processes

24 **5.1 Causation**
Traditional iron production and
pyroelectrolysis
Expressing cause and effect

25 **5.2 Stages (1)**
Traditional iron production and FINEX
steelmaking
Choosing appropriately between active and
passive

26 **5.3 Stages (2)**
Description of aluminium smelting
Describing functions of parts used in smelting

Unit 6 Planning

28 **6.1 Risk**
Article on the possible effects of earthquakes
Expressing likelihood

29 **6.2 Crisis**
Plans for dealing with blowouts
Future passive forms

30 **6.3 Project**
Plans for a nuclear power plant programme
Risk assessment
Future passive forms; comparatives

32 **Review Unit C**

Unit 7 Developments

34 **7.1 Progress**
Product evaluation of a handheld computer
Noun phrases
Revision: tenses and language forms

35 **7.2 Comparison**
Evaluating two types of touch screen
Comparing and contrasting

36 **7.3 Product**
System description of fibre optics
Technical versus non-technical terms

Unit 8 Incidents

38 **8.1 Missing**
Warehousing procedures
News story
Modal + perfect active and passive: must
have been + past participle

39 **8.2 Confidential**
Article on computer worms
Indirect questions

40 **8.3 Danger**
Story of a water pollution incident
Word building: noun, adjective, verb
Certainty in responses: to some extent,
definitely not

42 **Review Unit D**

Unit 9 Agreements

44 **9.1 Proposals**
Port security systems project
Security devices and definitions
Proposals and suggestions

45 **9.2 Definitions**
Basic and expanded definitions
Generic terms used in definitions: process,
instrument

46 **9.3 Contracts**
Clauses in contracts
Contract terms: shall, should (= if), otherwise
Conditionals: on condition that, provided that,
even if

Unit 10 Testing

48 **10.1 Test plans**
Seismic test reports
Hyphenated phrases acting as adjectives:
present/past participial phrases
Concision

49 **10.2 Test reports**
Sections in a test report
Expressions of damage, metal fatigue,
failure: corroded, buckled

50 **10.3 Test methods**
Non-destructive test methods
Language for chairing a meeting: Let's
start the meeting now. That concludes our
meeting

52 **Review Unit E**

Unit 11 Accidents

54 **11.1 Investigation**
News story of a disaster
Conclusions and recommendations: should/
shouldn't + passive

55 **11.2 Report**
Accident investigation report
Third conditional

56 **11.3 Communication**
Guidelines for effective oral communication
Phrases for assertiveness training: Can we
discuss our action plan now?

Unit 12 Evaluation

58 **12.1 Projects**
Evaluation report of a Compressed Air
Energy Storage system
Range of forms expressing necessity and
purpose

59 **12.2 Performance**
Appraisal interview
Phrasal verbs

60 **12.3 Innovation**
Nanotechnology
Material properties
Revision: could for present and potential
applications

62 **Review Unit F**

64 **Audioscript**

70 **Answer key**

1 Eureka!

1 Complete these texts, using the correct tense and the active or passive form of the verbs in brackets.

1 In 1968, Dr Spencer Silver (1) ___was working___ (work) at the 3M company in the US as a research chemist. He (2) _____ (develop) and testing different kinds of adhesives. By accident, he (3) _____ (discover) a 'low-tack' reusable pressure sensitive adhesive, in other words, one which (4) _____ (require) only gentle pressure to make one surface stick to another. Furthermore, the two surfaces (5) _____ (can separate) easily and then stuck together again. In 1977 3M (6) _____ (launch) the product as the 'Post-it®' note, and by 1980 it (7) _____ (sell) the product successfully across the USA.

2 In 1950, John Hopps (1) ___was doing___ (do) electrical engineering research in the field of medicine. He (2) _____ (work) with radio frequencies and their ability to raise body temperature. One day, he (3) _____ (discover) by chance that it was possible to restart a heart by electrical methods. The first portable pacemaker (4) _____ (invent) in 1957 and began (5) _____ (use) all over the world. In due course, the size (6) _____ (reduce), and the weight and size of the batteries (7) _____ (shrink), so that the pacemaker (8) _____ (can implant) inside the body.

2 Complete these dialogues, using the words in brackets and underlining the correct words. Use the present perfect continuous for the first question.

1 A: How long *has your company been extracting oil and gas from the North Sea?* (your company / extract / oil and gas / North Sea)

B: *For* / *Since* 40 years.

A: _____ ?
(how / crude oil / get to / refineries / shore)

B: _____ , but now it's brought ashore by pipeline.
(before / transport / tankers).

2 A: How long _____ ? (companies / produce / LPG*)

B: *For* / *Since* 1912.

A: When _____ ?
(motorists / start / use / LPG / cars)

B: _____ as an alternative to petrol and diesel.
(start / use / 1940s)

3 A: _____ ?
(sales / LPG / Europe / increase / decrease / recently)

B: *For* / *Since* the past eight years, _____ .
(sales / rise / steadily)

A: In your opinion, _____ ?
(what / cause / increase / sales)

B: These days, more LPG _____ .
(use / rural areas / heating / electricity generation)

*LPG = liquid petroleum gas

2 Smart drilling

1 Complete this article with the adjectives in the box.

> absorbent concentrated conventional flammable geological innovative liquid
> organic partial solid sufficient

Crude oil

Formation
Today's reserves of hydrocarbons (oil and gas) derive from ancient (1) ___organic___ materials like dead leaves and trees that settled on the bottom of seas and lakes. Heat and pressure combined to transform the organic matter first into kerogen, found in oil shales throughout the world, and in due course into (2) _____ and gaseous hydrocarbons.

Reservoirs
Oil reservoirs form only in certain (3) _____ conditions: the hydrocarbon materials must be buried so deep that pressure and heat combine to cook it; the rock system must be (4) _____ so that it acts as a reservoir; finally, as oil and gas tend to rise, there must be a cap of (5) _____ rock to prevent the crude oil and gas escaping to the surface.

Fold (structural trap) Crude oil reservoir

Fault (structural trap) Crude oil reservoir

Extraction
In a (6) _____ well, underground pressure is (7) _____ to force the crude oil to the surface. In time, the pressure fades, and pumps are required to raise the oil to the surface. When this method fails, water is pumped into the underground reservoir to force out the oil. In wells that are becoming depleted, more (8) _____ methods may have to be used, such as forcing steam down into the reservoir.

Oil shales versus oil sands
Oil shales are relatively hard rocks, called 'marls', that have not been exposed to enough heat and pressure to transform them into crude oil. The kerogen trapped in these marls is (9) _____, however, and can be transformed into crude oil by heat and pressure. Oil sands, on the other hand, are deposits of crude oil that have undergone a (10) _____ degradation. The resulting deposits form a heavy, (11) _____ form of crude oil, known as oil sands.

2 Join these pairs of sentences into single sentences. Start with the past participle in italics.

1 Oil shales are *classified* on the basis of their composition. They include carbonate-rich shales.

Classified on the basis of their composition, oil shales include carbonate-rich shales.
_____

2 The most common classification was *developed* between 1987 and 1991. It adapts terms from coal terminology.
_____

3 Oil shales are *described* as 'terrestrial' (earth) or 'marine' (sea). They are the result of the initial biomass deposit and its environment.
_____

4 Oil shales have been *used* since prehistoric times. They burn without any processing.
_____

5 Industrial mining was *started* in France in 1837. This was followed by further exploitation in Scotland and Germany.
_____

6 The mass production of cars was *accompanied* by an increase in petrol consumption. It helped to expand the European oil-shale industry before 1914.
_____

3 Lasers

1 🔊 02 Listen to a lecture about X-rays. Tick the topics that are mentioned.

☐ Discovery of X-rays	☐ How an X-ray machine works
☐ What is an X-ray?	☐ Using X-rays to examine metal beams
☐ Movement of electrons	☐ The use of 'contrast media' in X-ray technology
☐ Why do bones show up on X-rays?	☐ The use of lasers in medicine

2 🔊 03 Complete this description of an X-ray machine, using the words and phrases in the box and the diagram. Then listen again to what Ahmed said and check your answers.

> atoms   electric current   heat   tube   X-ray photons

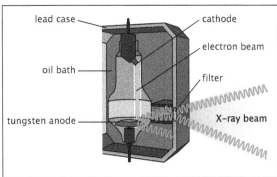

lead case
cathode
electron beam
oil bath
filter
tungsten anode
X-ray beam

In the middle of an X-ray machine, which sits in a
(1) _____*lead case*_____ , there is a (2) _____
at the top and a (3) _____ at the bottom.
An (4) _____ passes through
the cathode and heats it up. The anode at the bottom draws an
(5) _____ across the
(6) _____. This process releases extra energy in
the form of X-ray photons. This stream of high-energy photons
produces a lot of (7) _____ so there is an
(8) _____, which absorbs it and keeps the machine
from getting too hot. The lead case around the machine stops the (9) _____ from escaping in
all directions. Instead, they are emitted in a narrow (10) _____, which passes through a
(11) _____ and then goes through the patient. The patient's bones, but not the soft tissues,
appear on film as a negative image. This is because the bones have large (12) _____, which
absorb the X-rays, whereas the soft tissues are made up of small atoms.

3 Complete these parts of the lecture with the phrases in the box.

> Ahmed just explained   can now turn to   can see a diagram   going to talk about
> hand back   have to say now   Our aim   there are no   to ask Magda to take over
> to explain what an X-ray is   to talk about   like to take over   you have a look at this slide

1 ___*Our aim*___ in this short talk is, firstly, _____.
2 Now I'm going to ask Ahmed _____ the X-ray machine.
   Ahmed, would you _____?
3 If _____ on the screen, you _____ of
   an X-ray machine.
4 We _____ the next part of our presentation, and I'd like
   _____. She's _____ uses of X-ray
   technology in medicine.
5 As _____, soft body tissue doesn't show up clearly on
   an X-ray picture.
6 That's all I _____.
7 If _____ questions, I'll _____ to Hans.

4 Word list

VERBS	NOUNS	NOUN PHRASES	ADJECTIVES
absorb	bitumen	cathode ray generator	coherent
access	geology	flagship project	complicated
contaminate	kerosene	laser beam	concentrated
displace	petrochemicals (plural)	ruby crystal	conventional
emit	snake-well	synthetic crude oil	dormant
enable	**NOUNS**	**NOUNS**	dual
immerse	atom	bacteria	innovative
overlook	photon	fungus	isolated
threaten	spectrum	pocket	multiple
weave		syringe	neighbouring
		web	ongoing
		ADVERBS	partial
		back and forth	steerable
		instantly	swellable
		laterally	
		remotely	

1　Match the adjectives 1–9 with their opposites a–i.

1	_e_ coherent	a	unusual
2	___ complicated	b	distant
3	___ conventional	c	connected
4	___ dormant	d	total
5	___ dual	e	unclear
6	___ isolated	f	discontinued
7	___ neighbouring	g	active
8	___ ongoing	h	simple
9	___ partial	i	single

2　Describe how a laser works, using the highlighted words from the Word list in their correct form.

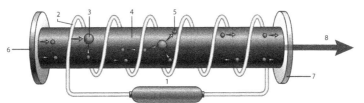

The power source is located below the (1) _ruby crystal_ and makes the tube flash on and off. Every time the tube flashes, the flashes inject energy into the crystal in the form of (2) _____, which are particles of light.

In the first stage, (3) _____ in the ruby crystal (4) _____ the energy from the light tube. When the atom absorbs a photon, it becomes more excited for a few milliseconds, and then returns to its original state and (5) _____ a new photon.

Next, the photons travel at the speed of light (6) _____ inside the ruby crystal. At the far end of the tube, a (7) _____ mirror reflects most of the photons, about 99% of them, back into the crystal, but lets a small number, about 1%, escape from the machine.

Finally, these escaping photons form a very (8) _____ beam of powerful laser light.

1 Spin-offs

1 Complete this text, using the correct tense and the active or passive form of the verbs in brackets. Underline the correct adverbs.

```
MBEF0132101(1.1)G
║║║║║║║║║║║║║║║║║║║║
```

A barcode is a machine-readable code that (1) _____ (use) to identify an object, person, document or collection of data. The code (2) _____ (present) in a series of parallel lines and spacings so that it can be read by an optical scanner, called a barcode reader.

(3) *Originally / First* prototype barcodes (4) _____ (use) to keep track of railway trucks, but the system (5) *repeatedly / mainly* ran into problems. (6) *Originally / Finally*, the system (7) _____ (take up) and developed by the grocery trade, and the first grocery barcode (8) _____ (scan) in 1974. (9) *Initially / Periodically*, take-up was slow, but by 1980 thousands of grocery stores in the USA were converting their systems to barcodes.

The UPC code (universal product code) is the one that is (10) *occasionally / generally* used today, and 11-digit barcodes identify every product with its unique identifying number. (11) *Regularly / Currently*, standardised barcodes (12) _____ (issue) by GSI, the most (13) *widely / normally* used supply chain standards system in the world.

Almost every grocery product has a UPC barcode. Barcodes (14) _____ (assist) in stocktaking and reordering, both in the grocery trade and in those manufacturing industries where many different components (15) _____ (assemble). At work, employees (16) _____ (scan) a 'job barcode' when they start and finish each job and thus record the time spent on it. In hospitals, a patient barcode (17) _____ (permit) staff to access the full medical records of a patient. Barcodes (18) _____ (enable) companies to track items such as rental cars, mail and parcels. Barcodes (19) _____ (use) by airlines to issue boarding passes, and to control and track the movement of passengers and their luggage. Tickets for public performances (20) *repeatedly / normally* include printed barcodes so that staff can detect counterfeit (fake) tickets and duplicates.

2 Complete this interview between a supermarket manager (M) and a reporter (R). Use different ways to express use and function: *to* + infinitive / *for* + *-ing* / *that/which* + present simple.

R: I guess you use barcodes (1) _____*to check*_____ (check) the day's sales in your store.

M: Barcodes are routinely used (2) _____ (reorder) and stock control. But in addition, barcodes are a complete system (3) _____ (help) us run our stores more efficiently and reduce waste. For example, barcodes identify fast-selling goods (4) _____ (need) to be reordered. And we can identify slow-selling goods (5) _____ (order) less often.

R: And what do you do with this information?

M: Well, the positioning of goods within a store is important. So, fast-selling goods occupy good positions, and then there are slower-moving goods (6) _____ (place) in less favourable positions, e.g. low down, or in corners. Barcodes are mostly used (7) _____ (provide) information and cut costs. The grocery business is very competitive.

R: How else do you use barcodes in your business?

M: All suppliers attach to each container (8) _____ (send) to our depot a UID barcode, (9) _____ (identify) it.

R: Sorry, what's a UID?

M: ID is short for identity and UID is a Unique Identifying Number. When it arrives at the depot, the barcode is read and stored on a computer (10) _____ (ensure) that the supplier is paid.

2 Specifications

1 Complete this product description with the adjectives in the box. There may be more than one possible answer.

> adequate   adjustable   comparative   enclosed   heavy-duty   long-lasting
> maximum   relative   sealed   variable

The Sturdy 801 is the most powerful mobility scooter in our range.

If you want a (1) _heavy-duty_ scooter that will carry you effortlessly up inclines of 15%, this could be the machine for you. With a (2) _____ passenger load of 225 kg, it will get you to your destination in (3) _____ ease. The scooter is not (4) _____, and so is only suitable for use in dry weather. It has a (5) _____-speed throttle with a top speed of 13 kph and an operating range of 60 km. (6) _____ performance data have earned this model a 'Best Buy' recommendation.

The power system uses a (7) _____ 1.9 hp motor, with a 5-year warranty. The electric charge is held in two (8) _____ lead acid batteries, more than (9) _____ for journeys up to 60 km. The joystick controller is (10) _____ and suitable for either right-handed or left-handed operation.

2 Join these pairs of sentences into single sentences, using a present participle phrase (verb + -ing). Some sentences require the additional word in brackets.

Example: 1 *Inflatable 35 cm wheels provide a comfortable ride, allowing drivers to travel across rough, open ground.*

1 Inflatable 35 cm wheels provide a comfortable ride. They allow drivers to travel across rough, open ground.
2 The operating range of 60 km allows drivers to return. They do not need to recharge the batteries. (without)
3 The padded seat provides comfortable support. It enables drivers to swivel to the side in order to get off.
4 The simple multi-functional control panel is easy to use. It doesn't require physical strength. (without)
5 The powerful headlight lights up the road ahead. It ensures that the driver is seen from the front.
6 The scooter has front and rear storage baskets. They provide space for day bags and shopping.
7 The rear coil spring suspension provides a comfortable ride for users. They cross uneven ground. (when)
8 There are bright rear lights and reflectors. They enable the scooter to be seen from behind.

3 Properties

1   **04** Listen to a brainstorming session and answer these questions.

    1   Is the meeting about a) constructing a new building, or b) modernising a building?
    2   Are the discussions focused on a) the structure, or b) the interior decorations?
    3   Are the suggestions concerned with a) the design, or b) the use?
    4   Which two things are agreed on for further action?

2   **04** Complete these phrases for inviting suggestions, praising contributions and introducing ideas. Then listen again and check.

    1   Who'd like _____ _____ _____?
        Anna?
    2   _____ _____ on that?
    3   OK, _____ _____ on that.
    4   What _____ _____ _____, Basil?
    5   _____ _____ _____ we get an
        architect ...
    6   Right, _____ _____ sense.
    7   Any _____ _____ on that?

    8   _____ _____ interesting.
    9   I've got _____ _____, Craig.
   10 _____ _____ _____ solar panels on
       the new roof?
   11 That's a _____ _____!
   12 Why _____ _____ _____ an
       appointment ...?

3   **05** Listen to a technical presentation about the same building. Tick the boxes 1–8 of the properties of the four materials mentioned in the presentation.

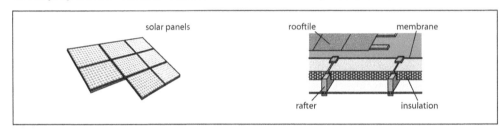

Properties of materials	Solar panel	Roof tile	Membrane	Insulation
Specification	*125 W* module conversion efficiency: _____	_____ tile	insulating membrane thickness: _____	thermal insulation board thickness: _____ (between rafters) _____ (below rafters)
1 durability (years)	☑ guarantee: _____	☐ _____ years	☐	☐
2 tensile strength	☐	☐	☐ same _____ and _____	☐
3 maintenance-free	☐	☐	☐	☐
4 strength:weight ratio	☐	☐	☐	☐
5 weight	☐	☐	☐ _____ per sq m	☐ _____ per sq m (thicker board) _____ per sq m (thinner board)
6 flexible	☐	☐	☐	☐
7 easy to cut	☐	☐	☐	☐
8 thermal protection	☐	☐	☐	☐

4   **05** Listen again and complete the notes in 3.

4 Word list

VERBS	NOUNS	COMPOUND NOUNS	ADJECTIVES
adjust	acoustic	bar coding	adequate
anchor	algorithm	daylight harvesting	adjustable
create	canteen	food court	ambient
deploy	compliance	joystick controller	applicable
detect	constraint	relative humidity	comparative
develop	counter-weight	robo-climber	cost-effective
enable	detonator	solar panel	enclosed
enclose	ergonomics (plural)	solar reflectance	incremental
generate	fabric	solar translucency	inflatable
install	feedback	stress sensor	long-lasting
jettison	incline	thermal protection	non-stick
occur	increment	ultra-violet ray	relative
optimise	landslide	web browser	sterile
protect	launcher	zero gravity	stunning
reflect	shield		toxic
utilise	spin-off		unrestricted
zip	Teflon®		variable
	treadmill		**ADVERBS**
	vapour		originally
	Velcro®		primarily
	weightlessness		

1 Underline the adjective that cannot be used with the noun on the right.

1	sterile	<u>inflatable</u>	long-lasting	**fabric**
2	flammable	solar	toxic	**vapour**
3	relative	adequate	cost-effective	**web browser**
4	variable	sterile	relative	**humidity**
5	adjustable	non-stick	long-lasting	**Teflon®**
6	adequate	applicable	incremental	**counter-weight**

2 Underline the noun that cannot be used with the adjective on the left.

1	**unrestricted**	use	<u>compliance</u>	access
2	**stunning**	design	architecture	detonator
3	**enclosed**	laboratory	Velcro®	cockpit
4	**cost-effective**	gravity	protection	daylight harvesting
5	**comparative**	humidity	translucency	browser
6	**ambient**	panel	temperature	humidity

3 `06` Write the words in the box on the correct lines. Then listen and check.

adequate   adjustable   ambient   applicable   comparative   enclosed   incremental
inflatable   relative   sterile   stunning   toxic   unrestricted   variable

1   1st syllable stressed: __*adequate*__ _____ _____ _____

_____ _____ _____

2   2nd syllable stressed: _____ _____ _____ _____

3   3rd syllable stressed: _____ _____

Section 1

1 Write questions and answers for a job interview for the position of Laser Applications Research Manager. Use information from the CV and these tenses: past simple, past continuous, present perfect simple, present perfect continuous.

Example: *1 Where did you do your degree?*

(From 1993 to 1996) I did a BSc degree in Physics at Imperial College.

	CV Details	Notes
1	1993–1996: BSc Physics, Imperial College	*where do degree?*
2	1996–1998: MSc Laser applications in materials and medicine	*what study MSc?*
3	1998–2002: PhD Laser physics. Research: crystal geometry and ion salts; atomic positions within crystals	*subject PhD?*
4	2002–2004: Durban University. Research: development of electron gun	*what doing?*
5	2004–2010: Institute of Laser Information Technologies (Senior Research Scientist)	*job title?*
6	2004 to present: Research: laser applications in medicine	*working in which areas?*
7	Work to date: Special tools and control systems for laser applications in knee surgery	*techniques developed?*
8	20 articles, mainly on laser applications in medicine	*articles written?*

2 Use these notes to form single sentences, using the past participle as a linking device. Start the sentence with the past participle if it is printed in italics.

Example: *1 The optical beam of a laser when applied to a small spot leads to intense heating.*

1 the optical beam of a laser, when it is applied to a small spot, leads to intense heating
2 lasers are *widely used* in manufacturing – they enable precise techniques to be used
3 lasers have many advantages + *compared* with mechanical approaches
4 lasers, which are *used* to drill fine, deep holes, bring high processing speeds and reduced manufacturing costs
5 medical applications include eye surgery and hair removal + related to outer parts of the body
6 a laser results in less bleeding and less cutting of tissue + *if it is used* for surgery,
7 optical fibre communication sends pulses of laser light along fibre optic tubes + it is *used* for long-distance data transmission
8 lasers enable precise position measurements to be made + they are *used* for range finding and navigation

3 Put these words into the correct order to make signposting sentences for a lecture.

Example: *1 I'd like to start by talking about lasers in manufacturing.*

1 lasers start manufacturing like in talking by about I'd to
2 photons see you escaping the slide will you look this at If
3 to I'm part of my move talk to the next on going
4 now Boris over take to ask I'll
5 partial you see this mirror it's a photo close-up in As can
6 points covered I've the main I think
7 Mona hand back Now to I'm to going
8 final move Let's talk of the section to the on
9 time of out run almost We've
10 a question anyone ask to like Would?

Section 2

1 Expand each set of notes into one single sentence. Link clauses using the present participle of the verbs printed in italics.

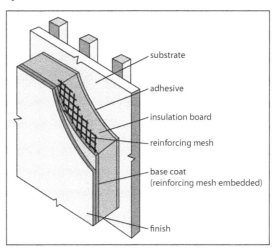

Example: *1 Exterior Insulation and Finishing Systems (EIFS) are multi-layer wall systems that are used on the exteriors of buildings to provide extra insulation.*

1 Exterior Insulation and Finishing Systems (EIFS) / multi-layer wall systems / used / exteriors of buildings / provide extra insulation
2 EIFS / be / excellent systems / insulate / solid masonry walls / *maintain* / interior dimensions of the building / *reduce* / thermal loss
3 EIFS / used / decades / make buildings more energy efficient / *reduce* / heating bills / up to 30%
4 bottom layer / include / sheets of 4–5 mm foam plastic insulation / fix / substrate / attached / the exterior / mechanical fasteners
5 middle layer / *consist* / fibreglass mesh / cover / with a coat of cement-type adhesive / apply / to the face of the insulation
6 outer layer / called the 'finish' / coloured, textured, paint-like material / apply / with a flat spreader / can be / rough or smooth / appearance
7 be / normal / exterior walls / contain moisture / so some EIFS / design / let water / escape / *allow* / wall / dry out

2 Complete the product description with the adjectives in the box.

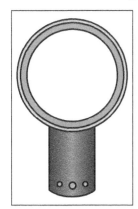

> adjustable ambient circular conventional easy fast-spinning incremental
> safe smooth sterile stunning uninterrupted variable

Bladeless fans with their (1) ___*stunning*___ designs have been around since 1981. They work differently from (2) _____ fans. Their technology draws air in and amplifies it – from 15 to 18 times, depending on the model. Without (3) _____ blades or a dusty grille that is far from (4) _____, they're (5) _____ and (6) _____ to clean.

The fan works by sucking in air through an inlet in the base and then forcing it out through a 1 mm (7) _____ outlet in the upper ring at a speed of about 80 kph. This produces an (8) _____ stream of smooth air, pulling extra air from behind the fan.

Controls are simple. Whereas conventional desk fans are top heavy and tend to overbalance, bladeless fans are easily (9) _____ and do not need clamping or unclamping. They have a (10) _____ speed control to suit the (11) _____ conditions. The (12) _____ airflow, which is not cut up by rotating blades, reaches a distance of five metres or more in a steady stream. (13) _____ increases in airflow do not reduce the fan's efficiency.

1 Product recall (1)

1 Rewrite this news story, changing all the verbs in the present continuous from the active to the passive. Make any necessary changes and use *by* + the agent where appropriate.

Begin: *The extent of the contamination is still being determined, …*

Northern Water has just announced that a contamination event has occurred in the northeast of the county, following the heavy rainstorms of the past few days. They are still determining the extent of the contamination, but are advising householders to boil their water until further notice. It is thought that the overflow system is still discharging sewage from the towns of Littlemore and Overmore into the rivers.

The Farmers' Cooperative Union is instructing farmers with properties adjacent to the River Bourne not to take water from the river for their farm animals, but to make other arrangements. Health officials are advising the public not to allow their dogs to enter the water, both along the River Bourne and along the Watership Canal. It is reported that the County Council is suspending all water sports, including the Canoe Regatta planned for 19 October.

Northern Water is currently employing extra inspectors to monitor water purity across the county, while the company is extracting water from the River Bourne to maintain supplies. The National Water Regulator is investigating the incident.

2 Complete this safety leaflet for householders, following the event reported in 1. Use each expression in the box once only.

> As a precaution   Even though   fails to comply with   for any inconvenience
> In the unlikely event that   In these unlikely circumstances   regrets to announce
> should be resumed   under certain conditions   will be subject to

Northern Water (1) *regrets to announce* that a contamination event occurred on 10th October. As a result, water purity currently (2) _____ the maximum safety levels of contaminants. The normal supply of safe drinking water (3) _____ within a fortnight. (4) _____, customers are advised to boil any water needed for drinking, cooking and the preparation of food until further notice. (5) _____ water supplies to the public become contaminated, further announcements will be made. Some streets in Littlemore may have their water supply temporarily suspended. (6) _____, water tankers will be placed in the street for customers to obtain drinking water. Northern Water apologises sincerely (7) _____ caused to its customers.

(8) _____ no complaints have been registered to date, any customers who wish to complain to the company should do so in writing. Claims for compensation will be considered (9) _____, although no guarantee of compensation can be made at this stage and all claims (10) _____ thorough investigation. Northern Water hopes to avoid similar problems caused by sewage discharge in the future.

2 Product recall (2)

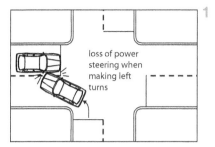

loss of power steering when making left turns

1 🔲 **07** Listen to a news report of a product recall. Write notes or underline the correct alternatives in the table.

1 Manufacturer	Quarry / <u>Murray</u> / Burridge
2 Model	Larissa / Surroba / Barrossa
3 Year of manufacture	_____
4 Problem	faulty _____ _____ _____ assembly
5 Action to be taken	_____

2 🔲 **08** Listen to an interview with a spokesperson for the company in 1. Write short answers for these questions.

1　What is the faulty part?
　the lower pinion bearing _____

2　Where is it located?
　_____

3　How could it malfunction?
　_____

4　What is the possible consequence?
　_____

5　Have there been any accidents or fatalities?
　_____

6　What have drivers noticed?
　_____

7　What repairs will dealers carry out?
　_____

8　What will be the cost for the owner?
　_____

3 Join these pairs of sentences into single sentences. Use *that/which* and a relative clause, or a present participial phrase. Include the modal verbs printed in italics.

Example: *1 The manufacturer stated that there were flaws in the engine valve springs, which could make the vehicle stall.*

2 The company issued a press release last week, covering vehicles from the 2008–2010 range.

1　The manufacturer stated that there were flaws in the engine valve springs. These *could* make the vehicle stall.
2　The company issued a press release last week. This covers vehicles from the 2008–2010 range.
3　Customers have complained about recent product recalls. These have caused a fall in resale values of used cars.
4　8,500 hybrid cars were recalled last year after tests revealed faulty fuel tanks. They caused fuel to spill after rear-end crashes.
5　Customers were being notified of the product recall by emails. These urged them to contact their local dealer without delay.
6　The air bag deployment signals are not reliable. This *could* result in the non-deployment of side air bags in frontal collisions.
7　The driver's side air bag inflator could come apart at the weld. This would prevent full inflation of the air bag.
8　Some drivers may notice extra vibration. They *should* bring the car to a dealer for service.

3 Controls

1 Read the explanations of two steering systems, and answer these questions.

1   What are the similarities between the autopilot and windvane systems?
2   What is the main difference in the way the two systems operate?
3   What is the function of the sensor in an autopilot?
4   In an autopilot, what stops the vessel from changing direction too frequently?
5   Which system steers a more precise course for a given compass bearing?

Autopilot

In a manual steering system, a crew member of a vessel moves the wheel in order to follow a compass bearing. The wheel is connected by a mechanical system to the vessel's rudder. An autopilot is a system where the crew member establishes details of the navigation, i.e. the compass direction to follow in advance, and then relinquishes control to the autopilot, which retains the instructions in its memory. If the crew member later wants to regain control of the vessel, he or she can override the autopilot at any time.

In a marine autopilot, a compass sensor controls an electric steering motor that is connected to the vessel's rudder by means of an actuator. Older systems turned the rudder every time the compass sensor registered a different bearing from the preset course. Today's autopilots are more advanced and include a feedback control. A 'deadband' refers to an area adjacent to the set course within which deviation is allowed and does not require a steering response. This prevents the autopilot from constantly steering the vessel back and forth across the desired course. The deadband can be varied for different conditions, e.g. rough or calm seas.

Windvane system

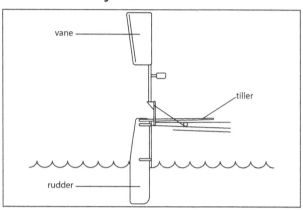

In a mechanical windvane self-steering system, the sensor is not a compass but a wind-sensitive vane. The crew member sets the windvane for a desired direction of sail. The windvane is connected by a mechanism to the boat's tiller, which moves the rudder at the stern of the boat. The windvane activates the steering mechanism and returns the boat to the desired point of sail (direction). A windvane self-steering system does not steer a constant compass course, but steers a constant direction of sail relative to the wind.

2   Find words in 1 with the same or similar meanings as the following.

Autopilot

1   a flat hinged surface at the back of a boat that is turned to steer it _____
2   to set or fix _____
3   to give away or hand over _____
4   to keep _____
5   to counteract or cancel _____
6   to notice or recognise _____

Windvane system

7   a flat metal or polycarbonate blade that turns in the wind _____
8   a long lever fastened directly to a boat's rudder to control the direction of the boat _____

4 Word list

NOUNS	VERBS	ADJECTIVES	PHRASES
actuator	comply (with)	additional	(not) in line with
autopilot	determine	economic	a split second
campaign	establish	military	subject to
compensation	experience	momentary	**COMPOUND NOUNS**
fix	interpret	pharmaceutical	combustion chamber
gearbox	kick in (start)	rare	cruise control
inconvenience	override	regenerative	drive-by-wire
input	panic	slight	wing surface
intervention	perceive	slippery	
negotiation	recall	top-level	
patch	regain	unlikely	
pharmacy	register		
precaution	relinquish		
probability	retain		
recall	skid		
rumble			
scare			
throttle			
update			

1 Underline the noun that cannot be used with the adjective on the left.

1	**economic**	situation	update	<u>gearbox</u>
2	**military**	campaign	recall	intervention
3	**top-level**	intervention	fix	negotiation
4	**slight**	delay	update	inconvenience
5	**slippery**	surface	patch	rumble
6	**momentary**	rumble	input	skid

2 Complete this text with the correct form of verbs from the Word list.

To (1) ___*comply*___ with training requirements, Mr Lo took Test CP503 on 7 May on the flight simulator. During the test flight, he (2) _____ a Category D incident. He (3) _____ that a warning light for the wing slats had come on. Instead of (4) _____ whether the wing slats or the warning light were faulty, he (5) _____ the autopilot, took control of the plane himself and brought the nose up. The plane climbed steeply, which scared Mr Lo. He then (6) _____ and went into a severe diving and climbing sequence, known as 'porpoising'. For three minutes he refused to (7) _____ control. After repeated warning signals, the autopilot (8) _____ and stabilised the plane. On landing, the plane (9) _____ on the slippery runway. Mr Lo (10) _____ control successfully and brought the plane to a standstill at the end of the runway.

3 ▶ 🔊 09 Listen to and repeat the nouns in column 1 of the word list. Underline the main stress on the words with more than one syllable.

1 Shutdown

1   **10** Listen to Part 1 of a talk about the Diamond Synchrotron. Are these statements *true* (T) or *false* (F)?

1   The Diamond Synchrotron is a particle accelerator that accelerates proton beams up to very high speeds.
2   A synchrotron is a particle accelerator that sends particles round in a circle at over the speed of light.
3   The LHC is designed to study particle collisions, whereas Diamond is designed to generate intense synchrotron light.
4   At Diamond, research is planned for drugs, diseases and the penetration of metal structures using X-rays, ultra-violet and infrared light, but this hasn't been carried out yet.

2   **11** Listen to Part 2 and complete the diagram with the phrases in the box.

> beamline   bending magnets   booster synchrotron   electron gun   linac   storage ring

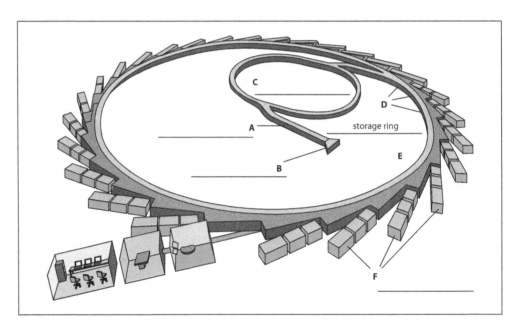

3   **12** Listen to Part 3 and complete the notes with the answers to the visitors' questions.

1   Electron beam produced: high-voltage cathode heated under ___*vacuum*___
2   Beam dimensions: _____ wide × _____ high
3   Electron beam in vacuum to avoid electrons colliding with _____
4   Vacuum: _____ times lower than atmospheric pressure
5   Electrons in storage: _____ hours
6   New electrons added _____ a day
7   Ideal beam: _____; produces light with _____ brightness
8   Synchrotron light: emitted when beam of electrons moves at close to the _____ and is bent by a _____

2 Overhaul

1 Read this story of an accident and number the pictures in the correct order.

A □ 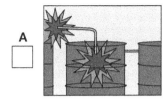 B □  C □  D □

Three men were killed and a fourth was seriously injured when flammable vapours ignited, causing two storage tanks to (1) *blow up*. A team of contractors had been engaged to empty a storage tank and (2) *put in* a new pipe between two of the site's four storage tanks. The team first (3) *cleaned out* a large storage tank, and then (4) *checked it out* for flammable vapours by a dangerous non-standard procedure: the foreman lit a welding torch and pushed it through a door in the tank. As there was no flash, he thought it safe to (5) *carry out* the welding job.

However, crude oil had remained in one of the other tanks, and the midday sun was causing the oil to (6) *warm up*. Vapours flowed into the adjacent tank and

escaped though an uncapped ventilation pipe just two metres from the site of the welding. Sparks from the welding torch rained down and ignited the flammable vapour. The flames (7) *flashed back* into the storage tank, causing a violent explosion, which (8) *blew off* the lid of the tank, as well as smashing the ladder and injuring the four men. Flames then (9) *shot through* the connecting pipe into the third tank. This caused an even bigger explosion.

Accident investigators warned of the potential (10) *build-up* of vapours when doing any hot work, especially grinding, cutting and welding at oil installations.

2 Match the words and phrases 1–10 in the article in 1 with these words and phrases with the same or similar meaning.

___ concentration
___ do
___ emptied
___ examined it
_1_ explode

___ get hotter
___ install
___ passed quickly through
___ removed
___ travelled quickly back

3 Complete these instructions for hot work, using the phrasal verbs in the box.

| carry on | carry out | clean out | look into | put off | rule out | start up | undergo | warm up |

1 Avoid hot work whenever possible and ___*look into*___ other methods such as cold or hydraulic cutting.
2 Always _____ a risk assessment before starting any hot work.
3 Test the area where hot work is planned, in order to _____ all possible sources of flammables.
4 _____ tanks before starting hot work, removing flammables like petroleum products or dust.
5 If vapour monitoring instruments are not to hand, _____ hot work until they are available.
6 Remember that midday temperatures can cause flammables to _____, vaporise and explode. So _____ monitoring throughout the day, especially if you _____ work again after a meal break.
7 Make sure that all operators involved in hot work _____ training and know how to use vapour-monitoring equipment.

3 Demonstration

1 Complete this description of welding processes, using the words in the illustrations.

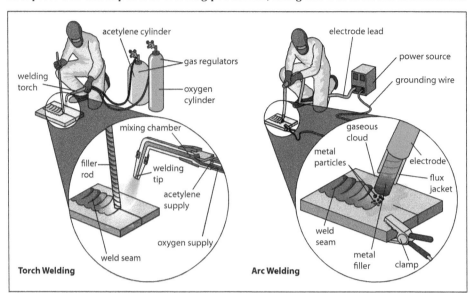

Torch Welding   **Arc Welding**

In welding, the first thing to do is wipe the work pieces clean of oil, dirt and dust. It is important, as seen in the illustration of **arc welding**, to protect the welding site from the effects of oxygen. For this reason, the filler rod, or (1) _metal filler_, is coated in a (2) _____. This serves to create a gas shield or (3) _____ around the welding site, which prevents it from oxidising and weakening.

Torch welding, e.g. with oxy-acetylene, is frequently used for maintenance and repair work. The process uses the (4) _____ of an oxy-acetylene torch to melt the work pieces and the (5) _____. Separate supplies of oxygen and (6) _____ are combined in the (7) _____ of the torch. The welder has to control the torch and rod simultaneously.

With **arc welding**, a (8) _____ from the (9) _____ is attached to the welding material by means of a strong (10) _____. Another wire, called the (11) _____, runs from the power source to the electrode, which is placed on the material that is going to be welded. When the (12) _____ is lifted from the material, an electric arc is generated that flashes between the two. The arc then melts the two work pieces together, along with the (13) _____ that helps to join them together.

Feeding the filler into the welding joint is a tricky job, because the welder has to do two things at once. As the metal filler melts, the welder has to feed the molten filler into the joint using small back and forth movements. At the end of the job, the (14) _____ is marked with small ridges, but should not protrude.

2 Read part of an audioscript of a welding demonstration. Put phrases a–f in their correct positions 1–6. Refer to the technical details in 1.

a) should not protrude
b) feed it into the joint with small back and forth movements
c) a cylinder of oxygen and a cylinder of acetylene
d) up from the work piece
e) Feed in the filler rod gradually while the torch melts it
f) should be attached to the welding material with a clamp

Now, for our next demonstration, which is arc welding. We're going to connect up the two leads from the power source. The green lead goes to the electrode lead. And the grounding wire (1) ___f___. Now lower the electrode down onto the welding site. OK? Now lift the electrode (2) _____ until it flashes. We've got the metal filler ready, haven't we? So, at the same time as we're getting the arc to flash, take the metal filler in the other hand and (3) _____.

Now it's time to do some oxy-acetylene welding. For this you need (4) _____. Join the hoses up to the torch and ignite the welding tip. (5) _____. At the end of the welding job, the weld seam (6) _____ from the welded work piece.

4 Word list

PHRASAL VERBS	(NOUNS: VERB + PARTICLE)	VERBS	SOLDERING / WELDING
bring about	build-up	celebrate	NOUNS
build up	check-up	initiate	arc
carry out	clean-out	overcome	arc welding
put off	meltdown	puncture	clamp
roll out	roll-out	smash	coating
rule out	set-up	vaporise	foam pad
set off	shutdown	**ADJECTIVES**	insertion
shut down	start-up	enormous	joint
start up	warm-up	sub-atomic	metal filler
stick up	**NOUNS**	super-conducting	oxy-acetylene
take over	absolute zero	super-cool	printed circuit board
take up	detector	violent	soldering iron
(VERBS: PARTICLE + VERB)	electric arc	**ADVERBS**	torch welding
downgrade	garbage	simultaneously	weld seam
overhaul	helium	smoothly	**VERBS**
undergo	interconnection		oxidise
upgrade	magnet		protrude
(NOUNS: PARTICLE + VERB)	particle		solder
inlet	proton beam		trim
input	restraint		**ADJECTIVE**
outburst	spark		molten
outlet	status		
output	support		
upgrade			

1 Complete this text with the correct form of verbs from column 1 of the Word list.

A violent explosion at Oakeshott Power Plant occurred as a result of an operator not following the correct start-up procedure for a boiler after a routine overhaul.

Last month, the company (1) _carried out_ maintenance on two of its smaller boilers, and also (2) _____ the oldest oil-fuelled boiler, Boiler 237. A new operator was instructed to bring Boiler 237 back on stream. Instead of following the standard instructions step by step, which takes several hours, he thought he could (3) _____ the boiler faster by doing it manually.

Although the upper burners ignited, unburned oil poured from the lowest ring of oil guns, which had failed to ignite. Fuel and vapour (4) _____ at the bottom of the boiler and exploded.

This (5) _____ the fire alarm and the fire service immediately rushed to the site of the boiler and (6) _____ the fuel supply. A spokesperson said that they could not (7) _____ negligence, and the operator will certainly have to (8) _____ extra training. All planned maintenance has been (9) _____ until Boiler 237 has been repaired.

Section 1

1 Complete this text about satellite navigation systems, using the correct form of the verbs in the box.

> comply   establish   indicate   interpret   maintain   override   panic   recall   regain
> register   relinquish   skid

Do sat-nav devices cause accidents?

Insurance companies have been (1) _registering_ an increase in accidents caused by drivers obediently following the instructions of their sat-navs. Trucks have become firmly wedged under low bridges. Motorists have (2) _____, braked suddenly and (3) _____ into oncoming traffic when instructed to turn into a 'No Entry' street. And in the States one driver had a lucky escape after steering his car onto a railway track after the sat-nav had (4) _____ a left turn.

GPS (Global Positioning System) is made up of a group of satellites which communicate with your GPS navigation device to indicate your position. Once the device (5) _____ your location, it is able to plan driving directions for you. However, about 300,000 crashes were caused in the UK in a single year, and a further 1.5 million drivers, when questioned in a survey, (6) _____ making sudden direction changes when following sat-nav instructions. One expert explained the problem: 'Some people rely on their sat-navs too much and (7) _____ responsibility for their own route finding. Common sense should always (8) _____ a sat-nav instruction, especially when you can see that a sat-nav direction is unsafe, illegal, or plainly out-of-date.'

Another motoring journalist commented: 'People have lost the skill to decode and (9) _____ a map when planning a cross-country route. We need to train drivers and their passengers to (10) _____ this useful skill.' A road safety organisation urged drivers 'always to (11) _____ alertness when driving. Also, it's essential to (12) _____ with road safety laws, which can vary from country to country, for example concerning cell phone use when driving.'

2 Rewrite these sentences to express similar contrasts, using the words in brackets. Use linkers with a capital letter at the start of the sentence.

Example: *1 A small above-decks autopilot is inexpensive, but it is not recommended for an ocean-going boat.*

1 A small above-decks autopilot is inexpensive. However, it is not recommended for an ocean-going boat. (but)
2 It is ideal for light conditions, but will have difficulty steering in rough seas. (Though)
3 Above-decks autopilots are easy to fit. Below-decks ones need professional installation. (whereas)
4 An autopilot will steer a vessel automatically. It is still necessary to keep a 24-hour lookout on board. (but)
5 A solo sailor can utilise a windvane in order to sleep, though he or she needs to check the wind direction from time to time. (Nevertheless)
6 A basic windvane controls the rudder directly. However, a more advanced system controls the tiller by means of a servo-oar. (whereas)
7 Self-steering systems allow a boat to be sailed continuously. However, sailing times may be longer than when the boat is sailed manually. (although)
8 Although a windvane does not keep the boat sailing on the required compass bearing, it steers the boat safely at a preset angle to the wind. (Nevertheless)

Section 2

1 Read this news story and label the diagram, using five of the words or phrases in the box.

blowdown drum   control room   furnace   level indicator   storage tank   splitter tower
outflow valve   vapour cloud

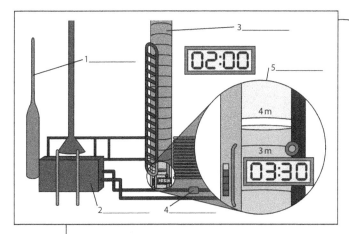

In 2005 at a US refinery, highly flammable petroleum components were released from a raffinate splitter tower and ignited, (1) <u>causing</u> violent explosions that killed 15 people and injured 170.

At 2 am, at the beginning of the refining process, operators introduced the feed (supply of petroleum) into the splitter tower through (2) *entry* pipes. The level indicator in the control room showed that liquid in the tower already (3) <u>reached</u> a height of two metres. However, this sensor didn't indicate levels above three metres. The tower over-filled, which set off an alarm at 3.30, but a second alarm failed to (4) <u>activate</u>. The intake of feed was halted, but a height of four metres had already been reached.

At 10.00, operators (5) <u>added</u> more feed to the splitter tower, which was already over-full. A valve controlling the outflow remained shut, contrary to operating procedures. The operators lit the burners in the furnace to (6) <u>warm</u> the feed, which was part of the normal process. The tower rapidly (7) <u>became fuller and fuller</u> and soon had 20 times more feed than normal. The feed may have reached a height of 42 metres, but the indicator showed less than three metres.

At 12.40, a high-pressure alarm was set off and two burners in the furnace were (8) <u>extinguished</u> to lower the temperature. As the automatic pressure valve didn't open, an operator used a manual valve to (9) <u>release</u> gases into the blowdown drum, from where gases were vented into the atmosphere. At 13.00, operators opened an (10) *outflow* valve to send liquid from the bottom of the tower to storage tanks. However, this liquid was extremely hot as it exited the tower, so the heat exchangers heated up the temperature of the feed going into the tower by more than 65 degrees Celsius. This liquid began to boil and expand, causing the level to (11) <u>rise</u> still further.

Ten minutes later, the liquid (12) <u>spilled over</u> from the top of the tower and down through vertical pipes. There was a great (13) *increase* in pressure on the emergency valves 50 metres below. The valves opened and liquid flowed into the blowdown drum. An alarm on the blowdown drum failed to go off and the drum filled up, releasing flammable gases and liquids into the air for about one minute. The liquid fell to the ground and vaporised, so that a large flammable vapour cloud (14) <u>formed and grew</u>. At 13.10, the vapour cloud exploded in a series of explosions. The ignition was probably (15) <u>caused</u> by a nearby diesel pick-up truck that (16) <u>turned on</u> its engine at that moment. The explosions triggered blast pressure waves, and this (17) *blast* of energy and heat killed and injured some of the workforce and resulted in considerable damage to the refinery.

All refinery operations were immediately (18) <u>halted</u>. In the next months, an investigation was (19) <u>conducted</u> to (20) <u>eliminate</u> the possibility of a similar disastrous sequence of events in the future.

2 Match the underlined verbs in the text in 1 with the phrasal verbs below with the same or similar meaning. Match the nouns in italics with the nouns below.

Phrasal verbs

___ brought about	___ filled up	___ let in	_1_ setting off
___ built up	___ go off	___ let out	___ shut down
___ carried out	___ go up	___ overflowed	___ started up
___ came up to	___ heat up	___ rule out	___ turned off

Nouns

___ build-up	___ inlet	___ outburst	___ outlet

Processes

1 Causation

1 Write sentences about iron production, using the notes and the words in brackets. Note the direction of the causation arrows (A → B = A causes B; A ← B = A is caused by B).

Example: *1 A supply of iron ore of uniform size and quality is a factor in iron production.*

1 supply of iron ore / uniform size and quality → iron production (factor)
2 some difficulties in production ← varying iron content / presence of sulphur and phosphorous / in ore (due)
3 in a blast furnace / introduction of super-heated air → reaction / where / burning coke / raise / temperature / 1,535 °C (give rise to)
4 coke / be / fuel / which ← heating coking coal without oxygen / temperatures of 1,000–3,000 °C / period / 18–24 hours (result)
5 carbon monoxide / release / from burning coke → reaction / iron ore (lead)
6 impurities like carbon and sulphur / remove / molten iron ← production process (result)
7 steady blast / hot air and gases / force / through / loaded raw materials → temperature increase / up to the melting point / iron (result)
8 blast furnaces / can operate / continuously / up to ten years ← their design (due)
9 iron ore / limestone / coke / continuously / charged / into / top of / blast furnace → continuous production process / without / loss / heat (result)
10 presence / limestone → impurities / gather / on the surface as slag / where / can easily / removed (cause)

2 Complete this article using the words and phrases in the box.

> anode by-product carbon carbon cathode cell electrolysis electrolyte
> emissions high-purity iron oxide oxygen sulphur

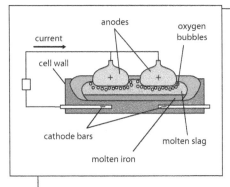

The ULCOS programme (Ultra Low CO$_2$ Steelmaking) is continuing to investigate steelmaking technologies that could reduce the CO$_2$ (1) _emissions_ of the industry by a large amount. (2) _____ is the sole technique for making aluminium and magnesium.

It is increasing its market share for the production of copper, zinc, and possibly other metals in future, like titanium. However, the future of pyroelectrolysis, the high-temperature production of steel, is still in doubt, as the (3) _____ reduction method is so efficient on a large scale.

Pyroelectrolysis uses electrical energy to do chemical work. (4) _____ is fed into an electrolysis (5) _____, where it dissolves in a solution of other molten oxides. An electric current is passed through the liquid (6) _____ in the cell from the top, the (7) _____, to the bottom, the (8) _____.

At the interface between the molten oxides and the cathode, pure liquid iron, or steel, is formed. The (9) _____ of the process, (10) _____, bubbles off the anode.

Unlike the production of aluminium and conventional steelmaking, no (11) _____ is involved. Currently, half a tonne of carbon is required to make one tonne of steel. Pyroelectrolysis could eliminate the processes that are currently necessary to remove (12) _____ (a contaminant) from steel, and this would enable companies to produce (13) _____ alloys more cheaply.

2 Stages (1)

1 Complete this text about traditional iron making, using the verbs in the box in either the active or passive form.

> burn contain cool create differ find make produce support use

Iron ore normally (1) __*contains*__ between 30–70% iron by weight. Ideally, iron ore should be a uniform size and have a high iron content with minimal amounts of sulphur, phosphorous and copper. High-grade iron ore like this (2) _____ in very few parts of the world. Another problem is that ores from different regions (3) _____ in quality, mineral content and form. In traditional iron making, this (4) _____ them impossible to use in their raw state. Iron pellets of uniform size and quality (5) _____ in special sintering plants.

The basic fuel that (6) _____ in traditional blast furnaces is coke. Coke (7) _____ by heating coking coal without oxygen at temperatures between 1,000 and 1,300 °C for 18 to 24 hours. At the end of this process, the coke (8) _____ by passing it through water sprays. Coke (9) _____ steadily and is able to (10) _____ the tremendous weight of the iron ore in the blast furnace.

2 ▶ 🔊 13 Listen to a DVD presentation about a new steelworks and complete the details.

Poco's new FINEX steelmaking plants.
1 Construction start date: __*August 2004*__
2 Inauguration date: _____
3 Location: Pohang, _____
4 Initial output: _____ million tonnes per annum (mta)
5 Proposed steelmaking plant: Orissa State, _____
6 Initial capacity (Phase 1): _____
7 Completion date (Phase 1): _____
8 Final capacity: _____

3 ▶ 🔊 14 Listen to an interview with the CEO of the steelworks. Write notes or underline the correct alternatives in the table.

	Traditional iron making	FINEX Process Technology
1 Process plant required: sintering plant (preparation of iron ores)	<u>Yes</u> / No	Yes / No
2 Process plant required: coking plant (preparation of coal)	<u>Yes</u> / No	Yes / No
3 Type of iron ore used	*iron ore pellets* Cost: 100%	_____ Cost: _____
4 Type of coal used	_____ Cost: 100%	_____ Cost: _____
5 Iron ore for Indian plant	N/A (not applicable)	Indian iron ore with high _____
6 Production costs	100%	_____% lower
7 Pollutants produced	sulphur, nitrogen oxide, CO_2: More / Less	sulphur, nitrogen oxide, CO_2: More / Less
8 Operational costs	100%	fewer _____ staff less _____ time

3 Stages (2)

1 Match the elements of the aluminium smelting process in the box with their functions 1–9.

> anode   carbon blocks   cathode   hopper   iron bar   ladle   molten electrolyte
> smelting pot   tap hole

1 _____hopper_____ a large container which holds the dry alumina above the pot

2 _____ a huge container with a steel outer shell where electrolysis takes place

3 _____ conductive blocks which line the bottom of the pot

4 _____ a cryolite liquid through which the electric current passes

5 _____ a conductive rectangular piece of metal that lies at the bottom of the pot

6 _____ the negative terminal, consisting of the carbon layer and iron bar

7 _____ the positive terminal at the top of the pot, consisting of two large pieces of carbon

8 _____ the opening at the bottom of the pot through which the molten metal is drawn off

9 _____ a heavy metal container which is used to transfer the molten metal to the casting house

2 Write questions and answers about the aluminium smelting process. Use the information in 1 to help you.

1 What __is aluminium made from__ ?
It's made from bauxite, which is mined in Africa, Australia and South America.

2 What is the main piece of equipment used for smelting aluminium and what is it made of?
_____

3 How much _____ ?
A smelting plant uses as much electricity as a large town.

4 How hot _____ ?
The electrolyte reaches a temperature of about 950 °C.

5 Where does the molten metal end up inside the pot?
_____

6 How _____ ?
In some plants it's siphoned off. In others it's drawn off through the tap hole at the bottom.

7 Can _____ be halted?
No. If smelting stops, the aluminium hardens. And the melting point of pure alumina is 2,054 °C.

8 How often _____ ?
The carbon blocks in the anode have to be replaced every 20 days.

9 Which _____ ?
Oxygen and fluorine are produced during the smelting process.

10 What _____ ?
It's non-magnetic, doesn't produce sparks, resists corrosion, and conducts heat and electricity well.

4 Word list

VERBS	ADJECTIVES	NOUNS	COMPOUND NOUNS
agitate	brittle	converter	aluminium oxide
calcinate	consecutive	coolant	blast furnace
charge	high-carbon	crusher	BOS furnace
dissolve	high-purity	crust	charging aisle
leak	inert	electrolysis	conveyor belt
pressurise	pure	electrolyte	digester tank
pulverise	residual	filtration	end product
siphon off	stressful	flux	fishbone diagram
smelt	waste	ladle	hot metal
suspend	**NOUNS**	lime	iron ore
tap	alumina	molecule	iron oxide
tilt	bauxite	pot	oxygen lance
PHRASAL VERBS	blow	precipitation	precipitator tank
bring about	bond	shortage	raw material
make up	by-product	slag	scrap steel
	carbon	sub-lance	sodium hydroxide
	conversion	supervision	tap hole

1   Match the words and phrases in the columns in each table, to make sentences.

Example: *1 Inert gas pipes are used in a BOS converter to agitate the mixture.*

1

1 Inert gas pipes 2 An oxygen lance	is lowered into the converter is added to the converter	to bond with impurities. to measure the carbon content of the molten metal.
3 Lime	is removed from the converter	to heat the mixture and melt the scrap.
4 Residual slag 5 A sub-lance	are used in a BOS converter is used	and is recycled. to agitate the mixture.

2

1 The mineral bauxite 2 Rocks of bauxite	is dissolved are calcinated	in rotary furnaces. in sodium chloride in a digester tank.
3 Ground bauxite 4 Pure alumina crystals 5 The crystals	sink to the bottom is extracted are pulverised	from the ground. in a crusher. of the precipitator tanks.

2   🔊 15 Write the words in the box on the correct lines. Then listen and check.

aluminium converter coolant electrolysis electrolyte filtration hydroxide
molecule oxygen precipitation precipitator shortage supervision

1   1st syllable stressed: _____ _____ _____ _____
2   2nd syllable stressed: _____ _____ _____ _____
     _____
3   3rd syllable stressed: ___*aluminium*___ _____ _____
4   4th syllable stressed: _____

1 Risk

1 Complete this article with the expressions of likelihood in the box. Use each expression once only.

> a high possibility   are likely to be   could happen   is almost certain to be   is no doubt that
> likely to happen   no doubt that   now a slim chance that   strong likelihood that
> The probability that   There's a strong probability   virtually certain   would be able to

Earthquake disaster waiting to happen

When the next earthquake strikes the *impressive* but crowded city of Istanbul with its population of over ten million, the consequences (1) ___are likely to be___ unbelievably terrible. A civil engineer showed us the crowded streets and pointed out some hazards.
'Here we have the *critical* problem of too many poorly constructed buildings in a very confined space. It is (2) _____ that an earthquake at around 7 points on the Richter scale would result in maybe 40,000 deaths and 120,000 injured.
(3) _____ most of the deaths and damage will occur in areas of substandard housing is very high. Look at the *disastrous* destruction that occurred in Haiti in January 2010. There (4) _____ the same catastrophe (5) _____ here.'
The *potential* problem of collapsing buildings is matched by a considerable fire risk. '(6) _____ that in the event of an earthquake about 30,000 natural gas pipes could fracture, resulting in hundreds if not thousands of fires.

An expert predicts that the tragedy of Haiti (7) _____ exceeded in this century. 'It is (8) _____ in Karachi, or Kathmandu, or in Lima. In Tehran in Iran, there's (9) _____

that a serious earthquake as strong as Haiti's could cause as many as 30–40,000 deaths.' As a typical example of cities at risk, Istanbul has a long list of serious problems:

- *insignificant* planning controls
- upper floors added illegally to an existing structure
- upper floors overhanging pavements below
- use of substandard building materials, e.g. salty sea sand and scrap iron
- construction of buildings on sandy ground with *minimal* foundations
- poor supervision of construction projects by engineers.

In addition, the North Anatolian fault runs just south of the city. Nobody doubts that a disaster is *imminent*.

However, there have been *minor* improvements in building design, and some substandard buildings have been reinforced or replaced. Several thousand volunteers are ready to respond to a disaster. 49 neighbourhood teams have been set up, each with a container of emergency equipment: crowbars, generators, stretchers and other emergency gear.
'If an earthquake struck this city, there is a (10) _____ neither army nor civilian rescue teams (11) _____ reach some areas for three days or more. In that case, there's virtually (12) _____ many people would die before rescue teams arrived. However, there is (13) _____ a local volunteer team may be close at hand to provide help.'

2 Match the adjectives in italics in the article in 1 with these adjectives with the same or similar meaning.

1 inadequate _____
2 about to happen _____
3 breathtaking ___impressive___
4 catastrophic _____
5 likely _____
6 negligible _____
7 serious _____
8 slight _____

2 Crisis

1 Complete the crossword.

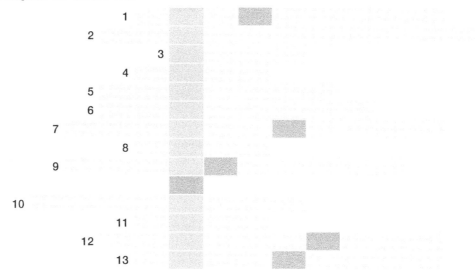

1 a thick patch of oil floating on the sea (2 words)
2 keeping something within limits so that it does not spread
3 to remove something from the surface
4 when liquids _____, they accidentally escape from a pipe or container
5 the top area of a quantity of liquid, e.g. the sea
6 a colourless, flammable liquid that prevents the formation of ice crystals
7 a well that is sunk to reduce underground pressure (2 words)
8 a floating barrier
9 the main pipe that carries oil up to the oil rig from the seabed (2 words)
10 a break
11 to insert under pressure
12 an underwater protective covering that is placed over the blowout preventer (2 words)
13 the top of an oil well where it exits the ground or seabed (2 words)

Vertical word: a pipe that is pushed inside a larger pipe or riser (2 words)

2 Change the verbs in italics in these sentences into the passive form.

Example: *1 A new task force is going to be set up to deal with future subsea oil spills in the Gulf.*

1 They *are going to set up* a new task force to deal with future subsea oil spills in the Gulf.
2 They *will make* underwater well containment equipment available to all oil and gas companies operating in the Gulf.
3 They *are going to develop* a new rapid-response oil spill containment system to help prevent another disaster like the 2010 blowout.
4 They *will make* the rapid-response system available for mobilisation within 24 hours.
5 They *will use* it on a range of equipment and in a variety of weather conditions.
6 By the end of this year, they *will have set up* the new operations centre in a permanent facility in Houston, Texas.
7 A spokesperson says that officials *are on the point of signing* contracts for capture vessels.
8 In addition, they *are about to order and deliver* a full range of manifolds, jumpers and risers to the operations centre.
9 In the meantime, they *will assess* existing equipment for use in the short term.
10 The new rapid-response project team are *going to retain* technical personnel with experience from the 2010 oil spill.

3 Project

1   **16**   Listen to an interview with a CEO about nuclear power plants (NPPs) in the Republic of South Africa (RSA). Answer the questions.

1   How many jobs will be created at the new plant? _____
2   Does South Africa use renewable energy resources? Yes / No
3   How many countries are building nuclear reactors at the moment? _____
4   Apart from South Africa and Asia, in which three places are nuclear reactors going to be constructed? _____
5   How is global nuclear capacity being extended?
    1) *building new plants*   2) _____
    3) _____

2   **16**   Listen to the interview again, and complete the table.

1	Medupi nuclear power plant (NPP)	Construction cost: 125 billion rand = $ _____ bn Power output: _____ MW
2	Future NPP programme	Number of reactors: _____
3	Future requirements	Need more _____ to be trained Need for support industry for _____, _____ and _____
4	Comparative costs of NPP Cost effectiveness	One NPP costs _____ more than a coal-fired plant Cheaper to build _____ NPPs than a single one
5	Suitability of NPPs for South Africa Future fuel supply	Locally produced _____ already available Plan to _____ in RSA for NPPs
6	Renewable energy programme	Generation of an extra _____ MW
7	Cost/benefits of NPP programme, compared with coal-fired plants	_____*Higher*_____ construction costs, _____ operating and maintenance costs, _____ fuel costs

3   **17**   Listen to a talk by an environmentalist, and answer the questions.

1   What are the benefits of the new Medupi NPP?
    (1)   _____*aesthetically stunning*_____
    (2)   close to the _____ of Medupi
    (3)   close to _____
    (4)   part of a _____
    (5)   construction of _____ for the
    _____

2   What does the risk assessment of the NPP disclose?
    (1)   catastrophic consequences for local population of
    _____
    (2)   serious possibility of _____

3   What is the most important environmental concern about the new NPP?
    (1)   imminent rise in _____
    (2)   NPP next to the sea _____

4 Word list

ADJECTIVES	NOUNS	COMPOUND NOUNS	OIL SPILL
aesthetic	agenda	bridge-cum-tunnel structure	**NOUNS**
breathtaking	avalanche		boom
calamitous	calamity	customs point	containment
catastrophic	concern	economic powerhouse	methanol
critical	consequence	land-reclamation	**COMPOUND NOUNS**
disastrous	consideration	manufacturing hub	insertion device/tube
imminent	environmentalist	risk reduction	marine biologist
insignificant	hurricane	tourist attraction	oil slick
man-made	likelihood	**VERBS**	relief well
minimal	magnitude	assess	remotely operated underwater vehicle (ROV)
minor	possibility	associate	
moderate	strategy	devise	
negligible	typhoon	encounter	risk assessment
potential		erupt	subsea cap
serious		extend	underwater robot
severe		occur	well head
slim		reclaim	**VERBS**
stunning		resist	insert
upbeat		result	leak
		voice	skim
			spread

1 Underline the noun that cannot be used with the adjective on the left.

1	**aesthetic**	beauty	consideration	<u>strategy</u>
2	**breathtaking**	consideration	tourist attraction	bridge
3	**calamitous**	avalanche	magnitude	typhoon
4	**imminent**	agenda	consequence	likelihood
5	**man-made**	containment boom	calamity	concern
6	**stunning**	consequence	project	structure
7	**upbeat**	assessment	recommendation	likelihood
8	**slim**	possibility	certainty	chance
9	**remote**	chance	probability	possibility
10	**high**	doubt	likelihood	probability

2 Complete this text, using compound nouns from the Word list in the singular or plural.

Following the explosion at the oil well last week, the company is taking all necessary action to contain and control the oil spill. First, it is trying to seal the (1) ____*well head*____ to stop the oil escaping. Next, it is trying to position a (2) _____ over the fractured riser. This will have the effect of capturing the oil spill so that the oil can be pumped to vessels on the surface. Thirdly, there are containment booms in place on the surface to restrict the (3) _____ so that it does not spread any further. In the next few days, an (4) _____ will be used to cut off the broken riser near the seabed. In addition, a (5) _____ will be activated to pump mud into the well. The (6) _____, presently under construction, will then be inserted into the broken riser. Drilling the (7) _____ to reduce the underground pressure is going to be started next week. Meanwhile, (8) _____ are already examining the effects of the oil spill along the coast. Finally, the government has ordered improved (9) _____ for all deep-water wells in future.

Section 1

1 Complete this news story with the words and phrases in the box.

> as a direct result of   As a result of   cause   caused   causing it to   direct cause of
> due to   gave rise to   leading   resulting in

In April 2007, 32 workers were killed and six were injured (1) _as a direct result of_ an accident at a steel works in China. Accident investigators found that the accident was (2) _____ by the incorrect use of substandard equipment.

At 7.45 local time, a ladle was being used to transport molten metal from the blast furnace to the ingot casting house. It was being positioned over a worktable in preparation for pouring, when it separated from the overhead rail. (3) _____ the collapse, all 30 tonnes of liquid metal, with a temperature of about 1,500 degrees Celsius, spilt. The liquid metal then burst through the windows and doors of an adjacent canteen where a work team was gathering to take over the shift. One worker explained the (4) _____ of the accident. 'As the ladle fell, it hit a flatbed in the ingot casting section, (5) _____ tilt and spill its contents.'

When the emergency services arrived, they were initially unable to provide assistance (6) _____ the high temperatures. After the area had cooled sufficiently, they rescued the six survivors, who were rushed to hospital.

Investigators concluded that the (7) _____ the tragedy was the use of a standard hoist instead of a heavy-duty hoist designed for dangerous smelting work. Additional factors that (8) _____ the accident were poor working conditions and inadequate safety measures. 'The work space was narrow, (9) _____ poor access,' said an expert, adding, 'Some firms have been expanding rapidly, (10) _____ to a neglect of basic safety.'

2 Rewrite this process description for aluminium smelting. Change the verbs in italics into the passive or keep them in the active, as appropriate. When you use the passive, decide whether or not to use *by* + the agent.

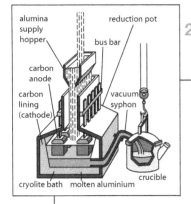

alumina supply hopper, reduction pot, bus bar, carbon anode, carbon lining (cathode), vacuum syphon, cryolite bath   molten aluminium   crucible

Every 20 days, they have to *replace* the anodes, as the two carbon blocks forming the anode have shrunk to one third of their original size. An operator uses a gantry crane to lift the anode assembly from the top of the pot and *place* it on a rack for cooling. Later they reuse the two three-metre rods that *hold* the carbon blocks. While the top of the pot is open, they *remove* waste matter from the crust with a pincer crane and *dump* it in a truck. Next, they *recharge* the pot from above with a fresh supply of alumina crystals. Finally, they *place* a new anode assembly on top of the pot, so that the smelting process can *continue* without a break.

Once a day, the operators *siphon* molten aluminium from the potlines. They *move* a collector tank into place to siphon off the molten metal. They *insert* a thick tube through a hole in the side of the pot, and *extract* air from the collector tank. This *creates* a vacuum, allowing them to *siphon off* the molten metal.

In the casting house, they *pour* some of the molten aluminium directly into moulds, where it slowly *cools* and *hardens*. The rest of the aluminium *passes* into furnaces where they *mix* it with other metals to form alloys. They *use* gases such as nitrogen and argon to separate impurities and bring them to the surface where they can *skim* them off. They then *pour* the resulting purified alloys into moulds and *cool* them rapidly. Water sprays *speed up* the cooling process.

Section 2

1 Complete this text about checklists with the adjectives in the box. There may be more than one possible answer. Ignore the alternatives in italics for now.

> breathtaking catastrophic critical imminent minimal potential serious
> significant upbeat

A US surgeon has reduced the number of (1) ___*serious*___ medical accidents by introducing a checklist strategy for hospitals. He is (2) _____ about the statistically (3) _____ fall in accident rates in hospitals that have adopted his strategy. 'There are two possible reasons why accidents happen. Either there is a (a) *slight / virtual* chance that our understanding of science and disasters is incomplete. Or there is a (b) *slim / strong* possibility that the knowledge exists, but is not applied.'

In hospitals, operating theatre teams take about 90 seconds to complete their checklists. Every single team member joins in going through the checklist. Previously, it was (c) *highly / almost* likely that a junior team member might question a procedure, but would be ignored. It is now realised that every team member contributes to the (4) _____ success of an operation, whether surgical or otherwise.

Once, pilots could fly a plane with (5) _____ checklists, because they understood and controlled all the plane's instruments and controls, most of which operated mechanically. But aviation systems developed so much that now a (6) _____ number of checklists has to be carried in every cockpit.

Accident investigators of three plane crashes pointed out the (7) _____ importance of junior crew members. In each accident, a junior had voiced concerns to the pilot, either to verify an order, or to check a landing approach that was (8) _____. In each case, the pilot ignored or rejected the question and went on to crash the plane with (9) _____ results. It is (d) *highly / virtually* certain that none of the crashes would have happened if the team had been following the checklist strategy.

2 Read the text in 1 again and underline the correct words in italics a–d.

3 Complete this text, using the correct passive future form of the words in brackets.

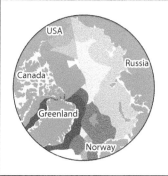

In August 2007, Russian scientists sent a submarine to the Arctic Ocean at 90° North to collect data in support of Russia's claim that the North Pole is part of the Russian continental shelf.

It is highly probable that the continental shelf limits claimed by Canada, Denmark and the USA (1) ___*are going to be defined*___ (going / define) legally over the next few years. In advance of the forthcoming international conference, a map and a set of briefing notes on the current situation (2) _____ (will / prepare). The map will identify known claims and agreed boundaries, plus areas that (3) _____ (going / claim) by more than one country. An agenda for the conference (4) _____ (point / draw up + discuss). Claims (5) _____ (about / file) at the UN (United Nations) by three countries. By the time of final agreement, scientific data about the outer limits of the continental shelves (6) _____ (will / collect + present) to the UN. It is expected that, as a result of the final conference, the Arctic seabed, which is rich in resources, (7) _____ (will / divide up) between the nations with claims to it, viz. Canada, Denmark, Norway, Russia and the USA.

All five countries bordering the Arctic Ocean expect that drilling in the Arctic region, with reserves of up to 100 billion tonnes of oil and gas, (8) _____ (going / carry out) in due course. Petroleum exploration (9) _____ (will / speed up) by the progressive melting of the Arctic ice cap, both along the Northwest Passage (north of Canada and the US), and along the Northeast Passage (north of Norway and Russia). Some experts predict that the Arctic region (10) _____ (will / make) completely ice-free by global warming by the year 2030.

1 Progress

1   Match the noun phrases 1–8 with the definitions a–h.

1	_g_ virtual reality	a) a component that can measure inclinations on two axes at the same time
2	___ augmented reality	b) a website that allows users to create their own profiles and view other people's profiles
3	___ digital value	c) a real environment with added computer-generated details
4	___ tilt sensor	d) a set of numbers generated by satellite positions that enables a location on the Earth's surface to be specified
5	___ digital compass	e) computer-generated information
6	___ social (networking) website	f) computer software designed to help users perform related tasks, e.g. graphics
7	___ GPS co-ordinates	g) a computer-generated environment that seems real
8	___ application (app)	h) an electronic navigational instrument that determines direction

2   Write a description of a prototype cordless handheld computer, using the notes in the table and the appropriate language forms.

The Voyager – handheld computing on the move

Aim	new handheld computers / develop / overcome / limitations of smartphone screen size
Components	large screen (25 mm diagonal) / built-in lithium-polymer battery / = 10 hours use-time; solid-state flash processor / choice of 16 GB/32 GB/64 GB capacity
Operation	features / available / inside a particular app; users / 'pinch, swipe and tap' / interactive screen / while navigating app / 12 mm-wide border / around / active display screen; useful / hold / Voyager / keep fingers off active display
Method	1 GHz ARM processor / includes integrated 3D graphics, audio, power management + storage → load apps + web pages from internet
Outdated technology	in the past / stylus / used / on interactive screen + handwriting recognition software; smartphones / have / disadvantage / very small screen
Recent development	virtual keyboard / provide / for Voyager → tap-typing / lots of text / much easier / than on smartphone
Work in progress	home screen interface / redesigned → incorporate user-controlled options; GPS live-data program / adapted → installed / as standard app / on Voyager
Further work required	Voyager / needs / fitted / camera → used / video conferencing + online chatting; this work / planned / next model
Future target	more people / turn to cordless handheld format / for / portable computing needs → number of available apps / no doubt / increase rapidly

Begin: *Aim: New cordless handheld computers are being developed to overcome the limitations of the smartphone screen size.*

Components: The system consists of a …

2 Comparison

1 Complete this text using the words in the box.

> capacitive capacitive clarity conductivity durability durable electrical
> light metallic resistive resistive resistive sharp up-to-date

There are two types of touch screen – the resistive type and the capacitive one. The
(1) ___resistive___ screen works like this: when you press the two metallic layers
together, you complete an (2) _____ circuit at the point of contact. By contrast,
the (3) _____ screen uses the (4) _____ of the user's finger instead
of pressure. The metallic layer on the screen stores an electric charge. Then the
finger draws current to the contact point. That's why, unlike the resistive screen, the
capacitive screen has only one (5) _____ layer, because the user's finger acts as
the other plate of the capacitor.

If the (6) _____ of the screen is important, that's a problem for the
(7) _____ type, because the two metallic layers on the screen filter out a lot of
light. By comparison, the clarity of the (8) _____ screen is much greater. That's
why resistive screens are only found these days on older computer monitors. Capacitive
screens are much more (9) _____.

The capacitive screen only works if you touch it with your finger. Conversely, a resistive
screen works if you use a pen or any other object that isn't (10) _____.
Concerning (11) _____, both types can be damaged if you scratch them
with a sharp object. If the screen is going to be in a public place rather than on a
desk, the (12) _____ one may be better as it is slightly more durable than the
capacitive one. However, compared with the capacitive screen, the resistive one can
wear out very quickly. As the resistive one works by pressing two layers together, that
automatically leads to wear and tear, and eventually the circuitry wears down. So in a
way the capacitive screen is rather more (13) _____, because the touch can be
very (14) _____ indeed and the screen will still respond.

2 Join these pairs of sentences about TVs into single sentences that have a similar
meaning. Use the words in brackets and make any other necessary changes.

1 Misawa is not like other manufacturers. It delivers top-class picture quality across
the price range. (Unlike)
Unlike other manufacturers, Misawa delivers top-class picture quality across the price range.

2 Misawa is not moving away from plasma. It believes that this format is best suited to
3D TV. (Instead of)

_____

3 Picture clarity may not improve this year. Its new TVs, however, will have greater
energy efficiency. (While)

_____

4 The outdated Sakatas are difficult to use. The new Sakatas are less difficult to use.
(much easier)

_____

5 The screen of the Sakata 101 is reflective. The screen of the Sakata 201 is slightly
more reflective. (than that)

_____

6 Expensive models have good acoustic quality. On cheaper models, it is poor.
(compared)

_____

7 You need a higher frame rate when watching sport. However, the software can
make the picture look unnatural. (although)

_____

3 Product

1 **🎧 18** Listen to an introductory talk about fibre optics. Are these statements *true* (T) or *false* (F)? Correct the false ones.

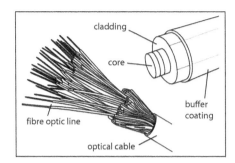
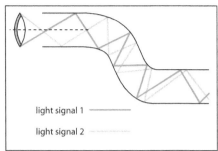

1 Fibre optics transmit light signals over long distances.
2 Fibre optics can transmit infrared light.
3 LEDs are more powerful but more variable light sources than lasers.
4 The cladding around a fibre optic line helps reflect the light inside the core.
5 Copper wires need fewer relay systems than fibre optics.

2 **🎧 18** Listen again and complete the speaker's words with the words and phrases in the box.

> another way of putting it is   or   that is   that is   that is to say
> to put that another way   to put this in everyday language

1 Now, I've got a slide here of a fibre optic line, _____*or*_____ 'fibre', as it's sometimes called. In the centre you have the core, _____ the hollow glass tube …
2 Fibres are very narrow; they can be from 9 microns to 62 microns in diameter. _____, they have the same diameter as a human hair.
3 … but they vary more with ambient changes, or _____, they are more variable in their performance …
4 … the light wave can travel great distances. _____ that this system allows signals to go round bends and travel a long way.
5 In due course, the light signal degrades, _____, it becomes weaker and less pure.
6 In addition, signals through copper wires suffered greater degradation, _____ there was often a loss of signal …

3 Match the words and phrases 1–9 in the text below with their less formal equivalents.

___ restored   ___ be greater than   ___ light flashes
_1_ greatly   ___ long-distance   ___ get rid of
___ impossible to decode ___ less pure   ___ shown

Scientists have demonstrated a system that could (1) *substantially* improve the data capacity of fibre optic networks. They say the growth in applications such as YouTube will eventually (2) *exceed* the limits of long-distance optical fibre links. The improved system would (3) *eliminate* most of the interference caused by other signals and amplifiers. Data is sent down an optical fibre as a sequence of (4) *bits coded into a light beam*, but it can become distorted over long distances. That can occur because of 'cross-talk' – signals sent down a fibre that influence one another. The result is that the digital signal becomes (5) *degraded* and (6) *unintelligible*. While this can be overcome with existing electronics, the result has the effect of reducing data capacity. Now, a team has (7) *demonstrated* a device that can clean up a signal and re-transmit it with fuller capacity. The signal for re-transmission is (8) *reconditioned* at the precise digital level required. This is of particular interest for (9) *long-haul* links, where data cables stretch across continents or oceans.

4 Word list

COMPOUND NOUNS	NOUNS	ADJECTIVES	VERBS
augmented reality	capability	acoustic	clap
camera lens	clarity	capacitive	label
centrifugal force	co-ordinates (plural)	conductive	oscillate
contact point	concept	durable	project
digital compass	conductivity	electromagnetic	protect
digital value	contaminant	hesitant	retrieve
disaster relief	durability	infrared (IR)	seal
live data	fluidity	interactive	solve
magnetic wave	gesture	metallic	squeeze
social network	limitation	nearby	superimpose
social website	overlay	outdated	vibrate
static data	projector	relevant	
surface acoustic wave (SAW)	radiation	resistive	
tilt sensor	reflector	specific	
virtual reality	resilience		
war zone	smartphone		
wave amplitude	stylus		
wireless broadband	substance		
	target		

1 Match noun phrases and nouns from columns 1 and 2 of the Word list with their definitions.

1 _____*stylus*_____: something shaped like a pen for writing on a computer screen or pad.

2 _____: a system of high-speed connection to the internet without using electrical or telephone wires.

3 _____: information that does not change from one moment to the next.

4 _____: a substance that makes something dirty.

5 _____: the place where you touch something, e.g. with your finger, or an instrument.

6 _____: a movement of part of your body, particularly hand, finger or head.

7 _____: the ability to last for a long time in good condition.

8 _____: a piece of equipment that makes a film or picture appear on a screen or flat surface.

9 _____: information that is constantly being updated.

10 _____: the ability to conduct electricity.

2 [▶ 🔊 19] Write the words in the box on the correct lines, according to the pronunciation of the 'a' vowels underlined. Then listen and check.

> c<u>a</u>pability cap<u>a</u>citive cl<u>a</u>rity cont<u>a</u>minant d<u>a</u>ta digit<u>a</u>l dis<u>a</u>ster l<u>a</u>bel limit<u>a</u>tion met<u>a</u>llic r<u>a</u>diation re<u>a</u>lity surf<u>a</u>ce

1 'a' as in 'say': _*capability*_ _____ _____ _____ _____

2 'a' as in 'cap': _____ _____ _____ _____ _____

3 neither 1 nor 2: _____ _____ _____

Incidents

1 Missing

1 Match the stages in the warehousing process 1–7 with the definitions a–g.

1 _f_ receiving	a) mechanised/automated/manual organising of goods, packets and cases
2 ___ put-away	b) moving goods to storage areas with pallet trucks, forklift trucks or automated systems
3 ___ storage	c) despatch of goods from the warehouse, usually by road
4 ___ picking	d) the assembly of products ordered by customers; items are pulled out from storage areas
5 ___ sortation	e) putting together shipments to make up the most economical shipping loads for specific geographical areas
6 ___ consolidation	f) checking incoming goods and verifying quantities
7 ___ shipping	g) keeping goods in a specific holding location until needed, while updating stock levels

2 Complete this news story, using the correct form of the words in the box.

> access   assumption   disable   incident   insider   inventories   outcome   pallets
> rule out   speculate   stress   thefts   threat   tighten up

Thieves cut a hole in a warehouse roof, slid down a rope and made off with $75 million in drugs over the weekend in one of the biggest (1) ___thefts___ of the year. The (2) _____ took place in Enfield. Once inside the warehouse, thieves moved (3) _____ of valuable pharmaceutical drugs to the despatch area. On Monday morning, staff discovered the break-in. After checking warehouse (4) _____, they discovered that the (5) _____ in terms of financial loss could be serious.

A police spokesperson expressed surprise that the thieves had been able to gain (6) _____ through the roof. He said that the burglars (7) _____ all the alarm systems and were possibly able to work over much of the weekend.

'We cannot (8) _____ the possibility that the thieves had information from an (9) _____, who could have told them where the drugs were stored. Our (10) _____ is that the thieves also knew the likely resale value of the drugs.'

A company spokesperson (11) _____ that security systems would be (12) _____ immediately. She also (13) _____ that many of the drugs would possibly be sold online. She said that the stolen drugs were a (14) _____ to the authorised supply chain and also to public health.

3 Change these sentences to give the same meaning, using the modal verb in brackets. Do not use the phrases in italics.

Example: *1 The theft could have been carried out by an insider.*

1 *It's possible that* the theft was carried out by an insider. (could)
2 *It's impossible that* a single person carried out the theft unaided. (can't)
3 *It's not possible that* the drugs were removed during the normal working day. (couldn't)
4 *It's virtually certain that* the movement sensor was disabled at the beginning. (must)
5 *It's wrong that* we stored the pallets of drugs in the low-security zone. (ought)
6 *One mistake is that* overhead security fittings were not routinely inspected. (should)
7 *It's recommended that* CCTV and security lights are installed for night-time use. (ought)
8 *I suggest that* all the security arrangements are reviewed and tightened up. (should)

2 Confidential

1 Read the text below about computer worms. Are these statements *true* (T) or *false* (F)? Correct the false ones.

1 If a computer worm manages to enter a computer system, it may not cause harm, since it will not change any control programs.
2 The Stuxnet worm is simple and smaller than average.
3 Computer personnel need to be careful when downloading programs from the internet, since these could contain a worm.
4 Worms can copy themselves, but cannot carry out unauthorised actions within a network.
5 The Stuxnet worm is responsible for the unintentional, wide but harmless distribution of copies of itself.

Companies have been warned of the dangers of (1) *downloading* programs from the internet after a Windows-specific computer worm called Stuxnet was discovered in 2010. (A computer worm is a kind of (2) *encrypted* software program that enters a computer network unnoticed inside another program, and then sends copies of itself to other computers on the network.)

It was first (3) *detected* by a computer firm from Belarus. 'Once it appears on a computer system, it (4) *triggers* an (5) *instantaneous* attack on control and monitoring programs,' a news agency reported. 'It is particularly suited to attacks on large industrial plants, but it can also (6) *wipe* files and programs. The software is complex and about half a megabyte in size, far bigger than most worms.'

Five manufacturing plants in Germany were found to be infected with the worm.

A (7) *corporate* spokesman for one of the companies issued a (8) *concise* statement that (9) *sensitive* company data was not at risk and that the worm was not active. 'However, we should warn that the worm can be passed on via the internet. All companies need to be very careful about (10) *authorising* only trained staff to download programs. Once present in a company's computer system, the worm is capable of (11) *executing* false commands. Since the whole of the Stuxnet code has not yet been (12) *decrypted*, its purpose is not yet known. It is capable of taking remote control of systems in a process plant, e.g. to override turbine RPM limits, or to shut down cooling systems, which of course would be disastrous in a power plant.' It appears that both the target and impact of potential disasters is highly important for the designer and distributor of the worm.

---

2 Match the words and phrases 1–12 in the text in 1 with these words/phrases with the same or similar meaning.

___ activates   ___ decoded   ___ noticed
___ carrying out   ___ delete   ___ permitting
___ company   ___ encoded   ___ private
1 copying   ___ immediate   ___ short

3 Rewrite these direct questions as indirect questions, using the words in brackets.

Example: *1 I would like to know how many attempts there have been to interfere with our company computer system.*

1 How many attempts have there been to interfere with our company computer system? (I / like / know)
2 Has anyone used the computer who has not been individually authorised? (please check)
3 Have any files been accessed or modified? (important / we / determine)
4 Who has been trying to obtain usernames or passwords of other users? (I / need / find out)
5 Has anyone copied software without permission or not? (we / got to / discover)
6 How much money has been charged for private use of computer services? (tell me)
7 Has anyone knowingly introduced a worm or harmful program into the computing facility? (vital / we determine)
8 Did anyone use the computing facility for a criminal act? (do / we / know)
9 Is anyone sending bulk emails from our computers? (we / got to / find out)
10 How many spot checks did we make last month and what did we discover? (please confirm)

3 Danger

1 Complete the vocabulary table with the missing words.

	VERBS	NOUNS	ADJECTIVES
1	emit	emission	
2	corrode		
3			flammable
4		hazard	
5			infectious
6		radioactivity	
7			reactive
8			toxic

2 🔊 20 Listen to a report of a water pollution incident. Write notes or underline the correct alternatives in the table.

1	Place	*Lowermoor Water Treatment Works*
2	Status of plant	*staffed / <u>unstaffed</u> / locked / unlocked*
3	Delivery	*aluminium oxide / aluminium sulphate*
4	Purpose of chemical	
5	Contents of tank	
6	People affected by incident	*10,000 / 20,000 / 30,000*
7	Concentration of aluminium in water	
8	Max. permitted concentration	

3 🔊 21 Listen to a question-and-answer session with a training officer about the incident in 2. Put a tick next to the questions he answers 'yes' to, and a cross if he answers 'no'.

1 Was the water supply badly contaminated as a result of the incident?
☑ *Without a doubt / 60,000 fish killed; soft mud in pipes stirred up*

2 Did the water authority warn consumers that the water was polluted and unsafe to drink?
☐ _____ / _____

3 In your opinion, was the water safe to drink?
☐ _____ / _____

4 I understand that the water authority did eventually warn customers of the contamination and advised them to boil their water before drinking it. Wasn't this sound advice?
☐ _____ / _____

5 Didn't the company have a programme of routine maintenance at their water treatment works?
☐ _____ / _____

6 In your opinion, was there a problem with management at that particular water treatment works?
☐ _____ / _____

4 🔊 21 Listen again. For each question in 3, write the words and phrases that the officer uses to introduce each answer, and note down the extra information she gives.

4 Word list

NOUNS	COMPOUND NOUNS	ADVERBIAL PHRASES	ADJECTIVES
assumption	hand scanning	event-based	concise
consolidation	memory stick	for the time being	corporate
hazard	scanning portal	on receipt	encrypted
incident	security alert	on sight	faulty
insider	security measure	on-demand	instantaneous
inventory	shrink-wrap film	out of action	preliminary
pallet	skull and crossbones	out-of-contact	relevant
picking	spot check	pre-scheduled	sensitive
portal	staging area	pre-set	unauthorised
receiving	storage rack	prior to	**VERBS**
shipping	**ABBREVIATIONS**	time-based	counter
sortation	CCTV (closed circuit television)	wake-up-and-wipe	disable
tag	RFID (radio frequency identification)		download
theft			execute
threat	SD memory card (secure digital memory card)		review
toxicity			rule out
	SMS (short message service)		speculate
	UU (unauthorised user)		stress
			tighten up
			trigger
			wipe

1 Match adverbial phrases from column 3 of the Word list with their definitions.

1 _out-of-contact_ : cannot be contacted because it is not connected to the internet
2 _____: dependent on a specified action being carried out
3 _____: dependent on the passage of a fixed period
4 _____: executed when it is requested
5 _____: makes the device switch on and remove all data
6 _____: fixed or planned beforehand
7 _____: planned beforehand to happen at a specific moment

2 **22** Write the words in the box in the correct column. Then listen and check.

assumption consolidation corporate encrypted incident insider instantaneous
inventory preliminary sortation toxicity unauthorised

1st syllable stressed	2nd syllable stressed	3rd syllable stressed	4th syllable stressed
	assumption		

Section 1

1   Write Sections 1 and 2 of a product comparison report, using these notes to compare two 3D TVs.

Begin: *1 Introduction: The purpose of this brief report is to compare …*

1 Introduction	2 Similarities
Purpose to compare 3D TV systems – Active Shutter 3D Display, Passive Shutter 3D Display – to help company choose best system for our hotels	• *Both types – 3D TV – send different picture to each eye. Brain combines them → one 3D image via use of active/passive shutter displays* • *Good clarity in both systems ← use top-of-the-range TVs. → Cost disadvantages for both systems.* • *Both systems show two images on screen. Both require glasses. For one system, glasses need power source, not for other.*

2   Write Sections 3 and 4 of the report, using these factsheets to compare the two TVs.

Begin: *3 Differences in technology: With the Active Shutter 3D display system, the TV …*
4 Differences in performance: These differences in the technology give rise to …

Active shutter 3D display	*Passive shutter 3D display*
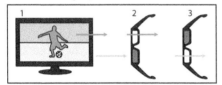	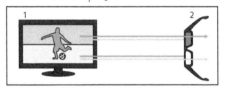
3 Differences in technology	4 Differences in performance
Active shutter • *TV shows the same two images, but not at the same time; TV flicks between showing a full-screen image for the left eye, and a full-screen image for the right eye* • *Glasses open and close shutters at the same speed→ each eye sees the correct picture* **Passive shutter** • *Two images shown on TV at same time* • *Each lens of polarised 3D glasses allows only one of the polarised images through, filtering out the other*	**Active shutter** • *3D glasses blink on and off; glasses need a power source → heavy!* • *Glasses must be suitable for own brand of TV* **Passive shutter** • *Glasses lighter, cheaper* • *BUT each eye sees only half the pixels on screen → 3D content won't be shown in full HD*

3   Complete Sections 5 and 6 of the report with the words and phrases in the box.

for these reasons   in addition   on the other hand   results in   since   whereas

5   Findings: 3D TV compared with 2D TV
  (1) ___Since___ the 3D screen is so huge (55-inch diagonal), it may be far too big for guest rooms.
  (2) _____, their size makes them ideal for suites. (3) _____ watching a 3D film at the cinema is great, the 3D effects on the TV were disappointing. (4) _____, it is necessary to sit close to the TV screen, which (5) _____ headaches and tired eyes.

6   Conclusion
  (6) _____, we consider 3D TVs more suitable for suites, and 2D HD sets better for guest rooms.

Section 2

1 Rewrite the numbered sentences 1–14 in this dialogue to give the same meaning. Use the modal verbs in brackets, and do not use the phrases in italics.

Example: *1 I can't understand what can have happened.*

A: (1) I can't understand what *has possibly* happened. (can) A lot of money has gone out of our company bank account in the past two days.

B: (2) *Is it possible that* someone has used your password? (might)

A: (3) No, *it's impossible that they* have done that, as I never write it down. (couldn't)

B: (4) *Is it possible that* they have interfered with an ATM in the high street? (could) (5) *It's possible* that somebody installed a camera to record people keying in their security numbers. (might)

A: (6) No, *it's virtually certain that* that hasn't happened. (can't) I never use the company card in an ATM for just that reason. Oh! I've just remembered. (7) I think I know what *probably* happened. (must) The tax office sent me an email and asked for the company's bank details. They said that they would send us a tax repayment of about €2,500. (8) *It's a mistake that* I replied, but I did. (shouldn't) (9) *It is highly likely that* it was a fake email. (must)

B: Bad luck! (10) *It was a mistake that* you replied, and *it would have been better to* have ignored the request. (ought / should)

A: (11) *Is it a good idea that* I tell the bank? (ought)

B: (12) Yes, *I suggest that* you go to your bank and explain what has happened. (should) (13) The money *will possibly* be retrieved, if you're quick. (might) (14) But you will *possibly* be unlucky and *it's possible that* you have lost the lot. (may / may)

2 Change the indirect questions into direct questions.

Example: *1 Why didn't you check the email properly?*

1 I would really like to know why you didn't check the email properly.
2 We must find out whether or not that email was from the tax office.
3 Please confirm when the money went out of the account.
4 It's vital that we determine whether anyone has ever written down their PIN or password.
5 We've got to discover whether or not she has been destroying all bills and receipts using the office shredder.
6 I need to find out if you have ever disclosed personal or security details in reply to an email.
7 Tell me if all the money was transferred from our Savings Account to our Current Account.
8 Please check whether he was using the same password for more than one account.
9 I wonder if anybody has replied to an urgent email asking them to confirm their bank details.
10 I'm not sure when we last had our anti-virus software updated to protect our computer network security.

1 Proposals

1 Match the devices 1–8 with the definitions a–h.

1 _f_ motion sensor	a) a device which detects the presence of a person in an unauthorised area and triggers an alarm
2 ___ thermostat	b) a network that repairs itself, and allows an alarm to be raised if one component device fails
3 ___ light sensor	c) a device that senses burning particles and triggers an alarm
4 ___ moisture detector	d) a battery-powered switch that 'wakes up' in order to transmit or receive a signal
5 ___ intrusion sensor	e) a device that detects a level of excessive humidity and raises an alarm
6 ___ radio-enabled switch	f) a device that detects movement in a room and triggers a light switch to turn on
7 ___ smoke detector	g) a device that triggers a light (circuit) and turns it on when the ambient light is insufficient
8 ___ self-healing mesh network	h) a device that measures temperature and triggers a switch to turn on or off

2 **23** Listen to a conversation between the security systems manager of a port and a security systems salesman. Answer these questions.

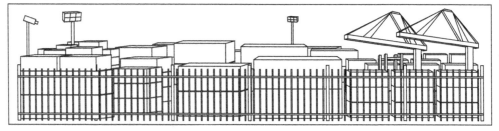

1 Is the manager interested in monitoring vehicles or people, or both?
2 How is security maintained at present?
3 What is the manager's reaction to the salesman's recommendations?
4 Is the new system designed to deal with current or future requirements, or both?
5 Has a detailed security proposal been submitted?

3 **23** Listen again and complete these questions and answers.

1 We need to ___*keep an eye on all vehicle movements*___ into and out of the port, as well as vehicles moving around inside it.
2 Incidentally, when _____ through the port, they often stop and ask for directions.
3 We need to _____ for these patrols.
4 How _____ a lighting and CCTV network?
5 The CCTV cameras must _____ inside and outside the port facilities.
6 I think we'll _____, including radio-enabled sensors.
7 It hasn't modernised, and it's _____, so …
8 So, as a first step, _____ for the systems we've discussed?

2 Definitions

1 Make definitions by numbering one phrase from each of columns 3 and 4. Some alternatives from column 3 are possible.

Name	be	Generic noun	Function
1 A thermostat 2 GNSS* 3 A tyre pressure gauge 4 A CO alarm 5 The ISO 6 Electroplating 7 A moisture meter 8 A carbon sensor	is	a detector ___ a device _1_ a gauge ___ an instrument ___ a process ___ a sensor ___ an organisation ___ a system ___	which measures the carbon content of molten metal. ___ for setting international standards. ___ which measures humidity in substances. ___ for measuring air pressure in vehicle tyres. ___ that provides global geo-spatial positions. ___ for measuring variations in temperature. _1_ that measures the concentration of CO. ___ that coats metal objects with another metal. ___

*Global Navigation Satellite System

2 Combine these notes with a basic definition from 1 to write an expanded, single-sentence definition. The devices are in a different order from 1.

Example: *1 A moisture meter is a portable battery-operated instrument with two electrodes which measures humidity in substances by measuring the electrical resistance between two points.*

1 portable, battery-operated / two electrodes / measures electrical resistance between two points
2 consists of / flattened thin-wall closed-end tube / connected / at open end / pipe / containing air pressure to be measured
3 voluntary / for trading categories of goods → allows / encourages / fair competition
4 navigational / uses GPS co-ordinates / means / electronic receivers
5 mounted on end of lance / in converter / during steelmaking process
6 small / battery-operated / contains electrochemical fuel cell / emits audible alarm / when concentration exceeds safe limit
7 plating / uses an electrical current / immerses them / bath / electrolyte
8 heat-sensitive, electrical / consists of two dissimilar wires

3 Find words in the puzzle to match the definitions below.

O	P	H	O	T	O	E	L	E	C	T	R	I	C
P	R	S	P	A	T	I	A	L	T	R	E	Q	O
M	E	Y	T	U	J	W	O	Z	P	M	L	S	M
F	C	D	I	S	S	I	M	I	L	A	R	B	P
U	I	Q	O	S	T	R	U	C	T	U	R	A	L
T	S	I	N	F	R	A	R	E	D	F	T	V	E
M	E	T	A	L	L	I	C	J	U	K	P	K	X
X	R	E	L	I	A	B	L	E	C	H	I	U	L

1 an invisible light that gives out heat _____
2 something which you can choose to do _____
3 relating to the size or shape of things _____
4 something that can be trusted _____
5 exact _____
6 made of metal _____
7 not exactly the same _____
8 using an electrical current that is controlled by light *photoelectric*
9 not simple _____
10 connected with the structure of something _____

3 Contracts

1   **24** Listen to a conversation between an airport executive and a contracts manager about the construction of a new terminal building. Are these statements *true* (T) or *false* (F)? Correct the false ones.

1   The contract has been discussed and there are no changes.
2   Two clauses about the system of payment have been changed.
3   The contractor will stop work on the project if they are not paid on time.
4   There is a disagreement between the contractor and client about quality standards.
5   In case of heavy rain and flooding, the contractor will pay for pumping out the water.

2   **24** Listen again and complete the conditions.

1   Site work must commence on the agreed date – *otherwise the contract will be cancelled* .
2   _____ , the penalty clauses for late completion will not be changed.
3   _____ is delayed, the penalty clauses for phased completion will stand.
4   Payment will be made monthly for the work certified as completed, _____ with the work.
5   The contract price is due on completion, _____ 2% is withheld for a period of 12 months …
6   Surveyors, building inspectors and the client can inspect the Works _____ the previous day.
7   _____ quality standards, an independent surveyor can be asked to assess the work …
8   De-watering of the site is allowed during heavy rain, _____ are informed at the start …

3   Match the legal/formal words 1–10 with the non-specialist words a–j.

1 _d_ nominate	a) beginning
2 ___ proviso	b) specify
3 ___ prior	c) previous
4 ___ commencement	d) choose
5 ___ intended	e) break
6 ___ unforeseen	f) proportionate
7 ___ stipulate	g) proposed
8 ___ breach	h) unexpected
9 ___ terminate	i) condition
10 ___ pro-rata	j) end

4 Word list

COMPOUND NOUNS	NOUNS	ADJECTIVES	VERBS
default mode	foil	dissimilar	arise
draft letter	intruder	heat-sensitive	bounce
economic downturn	loan	optional	draft
generic noun	photocell	passive infrared (PIR)	pinpoint
intrusion sensor	proposal	photoelectric	propose
mesh network	subsidy	prospective	struggle
mobile input device	theodolite	pyroelectric	tackle
motion detector	thermocouple	reliable	**(LEGAL/FORMAL)**
radio chip	tripod	scaleable	commence
radio-enabled sensor	voltmeter	self-healing	nominate
smoke detector	**(LEGAL/FORMAL)**	**(LEGAL/FORMAL)**	stipulate
strain gauge	breach	intended	submit
(LEGAL/FORMAL)	cancellation	prior	**VERB PHRASES**
legal action	circumstance	reasonable	cut down on
	commencement	unforeseen	draw up
	deadline		go about
	insurance		go for
	monies (pl)		home in on
	proviso		keep an eye on
			make one's way into
			put forward
			run down
			speed things up

1   Cross out the noun on each line that cannot be used with the verb phrase on the left.

		a)		b)		c)	
1	**cut down on**	a)	costs	b)	expenses	c)	~~difficulties~~
2	**draw up**	a)	a downturn	b)	a contract	c)	a proposal
3	**run down**	a)	supplies	b)	stocks	c)	insurance
4	**go for**	a)	an IT network	b)	a potential	c)	a package
5	**home in on**	a)	a specialist	b)	an intruder	c)	an area
6	**keep an eye on**	a)	security	b)	an area	c)	a presence
7	**make one's way into**	a)	a take-off	b)	a workshop	c)	a port
8	**put forward**	a)	a proposal	b)	a loan	c)	a suggestion

2   **25** Listen to and repeat these words from the Word List. Underline the syllable with the main stress in each word.

default, economic, generic, intrusion
cancellation, circumstance, commencement, insurance, proviso
dissimilar, optional, prospective, pyroelectric, reliable

1 Test plans

1 Complete the crossword.

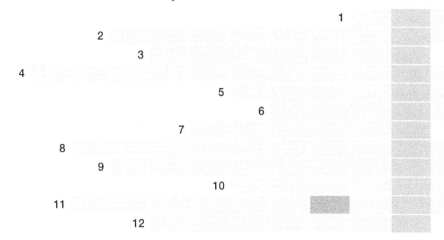

1   large, heavy and solid
2   to make a structure or a material stronger
3   produced by earthquake or similar artificial vibrations
4   the moving of something from its usual position
5   creating conditions that could exist in reality
6   to fix something deeply and firmly in something
7   a device that reduces shock
8   considerable
9   to cause to disappear, or disappear
10 the degree of an earthquake's force; size
11 a metal bar fixed to stabilise something (2 words)
12 to soften the effect of an impact

Vertical word: the rate at which the speed of something increases

2   Complete the more concise versions of these descriptions to give the same meaning.

1   The three tests of the series of experiments will analyse the cushioning effects of seismic dampers in buildings that are 15 storeys high.
    → The (1) _three-test_ series of experiments will analyse the cushioning effects of seismic dampers in (2) _____-_____ buildings.

2   Ten jacks, which weigh 50 tonnes, deliver a series of seismic simulations to the shake table that are carefully controlled.
    → Ten (3) _____-_____ jacks deliver a series of (4) _____-_____ seismic simulations to the shake table.

3   Houses that are easy to build, with wood frames, offer a cost-effective solution providing homes that resist earthquakes to a population which is growing fast.
    → (5) _____-_____ (6) _____-_____ houses offer a cost-effective solution providing (7) _____-_____ homes to a (8) _____-_____ population.

4   The research facility, which cost millions of dollars, is staffed by a project team that is highly respected, which has invested in equipment that saves labour and cuts costs.
    → The (9) _____-_____ research facility is staffed by a (10) _____-_____ project team, which has invested in (11) _____-_____ and (12) _____-_____ equipment.

2 Test reports

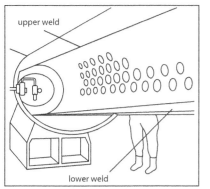

1 Answer these questions about the test report in 2, Section 2 of the Course Book, page 76. Paragraph letters (A–H) and section headings are provided for ease of reference.

Example: 1 A new type of load-bearing wall is needed which will not collapse during severe earthquakes.

1　C Introduction: Why has a new type of wall been designed and constructed?
2　G Test objective: What kind of sideways force was the experiment designed to test?
3　B Experimental setup: What was the difference between the two wall samples, and how was seismic movement measured?
4　A Procedure: What kind of experiment was set up, and were the two walls tested separately or together?
5　H Results: On which storey of the building was the greatest stress expected? How much damage was visually apparent after the tests?
6　F Conclusions: How great a seismic shock is the unreinforced wall expected to resist?
7　D Further testing: Which of the walls will be subjected to further tests, and how severe will the tests be?
8　E Appendix: How long did the earthquake simulation last, and at what magnitude?

2 Read paragraphs A–D below. Match them with four of these sections of a test report.

___ Introduction　　　　___ Experimental setup　　　___ Results
_A_ Test objective　　　___ Procedure　　　　　　___ Conclusions

A: The primary purpose of the test was to confirm whether the four remaining boilers had flaws in their materials. The second purpose was to discover the strength of welds in the repair work carried out on the boiler system.
B: Figures 1–10 in Appendix III give the main findings. Both non-destructive testing methods revealed serious flaws in three of the boilers and less significant flaws in the fourth boiler. Flaws were present both in the welds and also in materials that had not undergone welding repairs.
C: Each team followed a similar plan of action. Access was opened up by the shipping company, so operators could gain access to all interior and exterior parts of the boilers. Test data was collected from sensors for both methods and stored on computer. The data was analysed later.
D: On the test day, two teams were assigned to the boilers for visual inspections and non-destructive tests. One team carried out magnetic particle testing, while the other team used ultrasonic testing. The two teams had no contact with each other, and were instructed not to compare notes after the tests.

3 Match these two paragraphs with the remaining test report sections in 2. Underline the correct alternatives.

E: From these results it can be concluded that the explosion in Boiler 3 was the direct cause of *metal fatigue* / *design errors* along the deck *trusses* / *weld seams*. In addition, the condition of three of the remaining boilers is unsafe: they could be subject to similar *collapse* / *explosions*. Thirdly, the condition of these boilers is so *corroded* / *buckled* that repair is impossible and *replacement* / *structural support* is the only solution.
F: The *high-rise block* / *vessel* suffered a *catastrophic* / *faulty* fire after one of its five boilers exploded, killing two crew members and severely injuring four others. The accident happened while the ship was preparing to leave harbour and was therefore *bringing about* / *building up* maximum power. We were engaged by the Maritime Safety Board to carry out tests on the remaining *three* / *four* boilers. This is the report of the test, carried out in October last year.

3 Test methods

1 ▶ 🎵 **26** Listen to a discussion about non-destructive test methods for two projects. Complete the Method column of this table with the first and second choices for each project, as well as the unsuitable methods.

Project	Method	Reasons
Bridge	1st choice: _ultrasonic_	
	2nd choice: _____	
	Unsuitable: _eddy current_	_flaws may be too deep to detect_
	Unsuitable: _____	
High-rise apartment block	1st choice: _____	
	2nd choice: _____	
	Unsuitable: _____	
	Unsuitable: _____	

2 ▶ 🎵 **26** Listen again and complete the Reasons column of the table in 1. Note the reasons why methods are chosen, and why other methods are not chosen.

3 ▶ 🎵 **27** Complete as much as you can of the chairperson's language. Then listen to the chairperson's parts of the discussion and check your answers.

1 Let's _____ _start the meeting now_ _____. Right. The _____ is the Douro Bridge.

2 Carmen, could you _____ here? What did you _____?

3 Thank you, Carmen. That was _____. Mario, would you like to _____?

4 Before we come to a decision, I'll _____ so far.

5 It's time for us _____.

6 Let's _____ on our agenda.

7 That was _____. Carmen, would you like to _____?

8 Shall we _____ for discussion?

9 OK, we need to _____.

10 Then that _____. Excellent!

4 Word list

NOUNS	COMPOUND NOUNS	ADJECTIVES	VERBS
acceleration	anchor rod	destructive	assume
analysis	coil winding	excess(ive)	cushion
beam	eddy current testing	experimental	deduce
coil	kinetic energy	ferromagnetic	dissipate
damper	magnetic particle testing	linear	embed
displacement	push-over test	load-bearing	exceed
flaw	radiographic testing	long-running	fabricate
flux	shake table	long-term	induce
graph	shock absorber	massive	magnetise
magnitude	test probe	mid-rise	penetrate
objective	ultrasonic testing	non-destructive	plant
pedestrian		optical	reinforce
radiation		radiographic	simulate
setup		seismic	subject
simulation		significant	sway
strain		steel-reinforced	withstand
transducer		**ADVERBS**	yield
		apart	
		incrementally	
		instinctively	
		sideways	

1 Complete the text using the correct form of words from the Word list.

A mid-rise (1) ___experimental___ building has been attached to a huge
(2) s _____ _____ in Japan. The building is fitted with 240
(3) d _____, strain and (4) a _____ sensors to allow optical
monitoring by video cameras.

The building is fitted with seismic (5) d _____, which are capable of
absorbing much of the kinetic energy from the movement of the house. Engineers realise
that the house could suffer (6) m _____ damage, and even collapse: it is a
(7) d _____ test, as opposed to a (8) n _____ test.

The project team hope that the tests will yield (9) s _____ data about how
well the (10) s _____ dampers cushion the effects of the
(11) s _____ earthquakes.

The experiment is part of a (12) l _____ engineering project to design wood-
frame houses that can withstand powerful earthquakes.

Section 1

1 Match the non-specialist phrases in the box with the legal phrases in italics in the sentences below.

> as long as   before   except   except when something else is agreed
> if a disagreement occurs   in compliance with   in the case of   on condition that
> satisfactorily

1 The Employer will pay the Contractor the fixed sum of the Contract, *with the proviso that* additional sums may become payable as a result of variations in the Contract. _on condition that_

2 The Contractor will provide all labour, materials and equipment to carry out the Works *in accordance with* the plans provided at the start of the Contract. _____

3 The Contractor will carry out the Works *to the satisfaction of the Client* with reasonable care, skill and diligence. _____

4 *Unless otherwise agreed* in writing, the Employer agrees to obtain the necessary approvals and permits *prior to* the commencement of the Works. _____ / _____

5 The Contractor shall be granted an extension of time *in the event of* exceptionally bad weather. _____

6 *Should any dispute arise* under the terms of this agreement, both Parties agree to appoint an independent and impartial third party to help them settle the dispute. _____

7 The Contractor shall not be responsible for frost damage after completion of the Works, *other than* where the damage is caused by the negligence of the Contractor. _____

8 The Works are considered completed only on approval of the Employer, *provided that* the Employer's approval is not unreasonably withheld. _____

2 Complete the transcript below with the correct form of the verb phrases in the box.

> cut down on   follow up on   go about   go for   home in on   keep an eye on
> make our way towards   put forward

The CEO of Asia Cargo Handling is showing a visitor around the terminal.

Welcome! Good to meet you! Let's (1) _make our way towards_ the IT Suite. … Here we are. This is the hub of our cargo handling operations. When this terminal was designed, we (2) _____ an advanced mechanical cargo handling system and the latest IT. This has enabled us to (3) _____ all aspects of our operation. With our energy-saving programme, we've also been able to (4) _____ our energy consumption considerably. … About 95% of our operations are systematised. Whenever we have a problem, our network of radio-enabled sensors and CCTV cameras allows us to (5) _____ a zone and observe what's going on. Let's go into the meeting room. … Now, (6) _____ our phone conversation of last week, we would be very interested in (7) _____ some proposals for your bulk cargo handling and freight forwarding requirements. Our contracts department can (8) _____ drawing up a contract, should you decide to move on to the next stage.

Section 2

1 Refer to the four factsheets in Section 3 of the Course Book, page 79, and tick the boxes in this table to show what the testing methods can do.

		Eddy current	Magnetic particle	Ultrasonic	Radio-graphic
1	used only with ferromagnetic materials		✓		
2	used only with conductive materials				
3	used with any materials				
4	can detect, locate and measure flaws				
5	can detect and locate flaws				
6	can detect flaws at various depths				
7	can only detect flaws on or near the surface				
8	operator must have access above and below the flaw				
9	operator must have access above the flaw				
10	test instrument must contact one surface of the material				
11	test instrument must contact the surface above the flaw				

2 Write five sentences comparing the advantages and disadvantages of the four testing methods, using the answers to 1. Use *but, while, whereas* and *however* as necessary. Write one sentence for each of these groups of lines in the table: lines 1–3; 4–5; 6–7; 8–9; 10–11.

Begin: *1–3 Ultrasonic and radiographic testing methods can be used with all types of materials, whereas …*

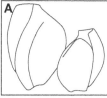

3 Read the descriptions 1–4 and match them with the pictures A–D.

1 The use of modelling software in three dimensions today allows designers to create forms that are extraordinarily complex. An external source of light which comes from an LED that uses a low level of energy throws light onto an element which is shaped like a leaf, which is made of PMMA that has been thermo-formed.

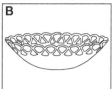

2 The chair is made entirely of polypropylene which has been strengthened with glass fibre. Nitrogen is injected at high pressure during the moulding process, and this reduces the length of the production cycle and also introduces an internal air cavity so that less material is required. The cycle time of three minutes demonstrates the impressive efficiency of injection moulding when assisted by gas.

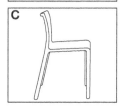

3 This range of items is produced by a machine that can be moved that extrudes strands of plastic that are like spaghetti. The bowl that is exhibited here was made of a polymeric resin that can be recycled and which was developed specifically for the extrusion of profiles where high clarity and good resilience are required.

4 The planters were designed by solving their complex geometry with digital modelling using three dimensions. The objects that have been coated with lacquer are made of polyethylene that has been through a rotational moulding process, which is like the one used in producing large rubbish containers.

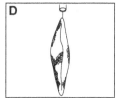

4 Rewrite each sentence in the descriptions in 3 as one concise sentence.
Begin: *1 The use of three dimensional modelling software today allows designers to create extraordinarily complex forms. An external, low-energy …*

1 Investigation

1 **28** Listen to a news report and answer these questions.

1 Where did the disaster happen? ___Kolontar___, _____

2 When exactly did it happen? _____

3 What were the consequences? _____ dead; _____ injured;
_____ evacuated

4 What were two possible causes of the accident? _____
_____

2 **28** Listen again and complete the notes in the table.

1	Area affected	_____7_____ villages and towns, _____40_____ sq km, _____ , _____
2	Plant	alumina plant, with _____ employees
3	Reservoir	_____ in area; storing _____ tonnes of sludge
4	Spillage	_____ m³
5	Counter-measures against further flooding	new _____ : _____ long, approx _____ high
6	Purpose of reservoir	storage of _____ sludge, a by-product of the alumina production process, while it _____
7	End-use of waste	dried red mud can be used for _____
8	Current use of waste	not _____ ; companies build another _____
9	Survey	took place _____ ago
10	Suitability of site	not suitable for reservoir, owing to _____

3 Rewrite the following sentences as conclusions and recommendations, using *should* or *shouldn't* and the passive form.

Example: *1 The reservoir should not have been constructed with an earth dam on an unsuitable site.*

Conclusions

1 They constructed the reservoir with an earth dam on an unsuitable site.

2 They didn't take into account the possibility of heavy rains when the capacity was determined.

3 They didn't periodically inspect the structure of the dam wall.

4 They didn't recycle the red mud after drying out.

Recommendations

5 They should take the company into public ownership for a period of two years.

6 The disaster agency should inspect all documents concerning the reservoir, in case negligence was involved.

7 Workers should wear protective masks and outerwear until further notice.

8 The disaster agency should continue to monitor the concentration of alumina dust in the air.

2 Report

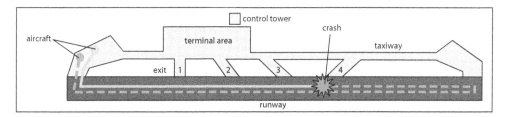

1 Look at the diagram of a plane crash and the events below. Number the first group of sentences a–h from 1–8 to show the order in which the events happened. Then number the second group i–p from 9–16 to show the correct order. Use the words and phrases printed in bold to help you.

Group 1: 1 _d_; 2 ___; 3 ___; 4 ___; 5 ___; 6 ___; 7 ___; 8 ___

Group 2: 9 _k_; 10 ___; 11 ___; 12 ___; 13 ___; 14 ___; 15 ___; 16 ___

Tenerife airport disaster

a) **Owing to the number of planes** delayed in the terminal area and on the taxiway, planes were forced to taxi along the runway to position themselves at the end for take-off.

b) **It was later found** that there were several communication misunderstandings between the controller and the pilots of both the KLM and Pan Am planes.

c) **A further delay** was caused by a KLM 747 deciding to refuel, which took an estimated 35 minutes.

d) **At 17.06 local time** on March 27 1977, two Boeing 747s collided on the runway of Los Rodeos Airport on the Canary Islands, with the loss of 583 lives. It remains the worst accident in aviation history.

e) **Shortly afterwards**, the Pan Am plane was also instructed to backtaxi, following the KLM plane.

f) **Once** extra flights started arriving at Los Rodeos, the airport quickly became very crowded.

g) **After that length of time**, the KLM plane was instructed to backtaxi to the end of the runway and await clearance for take-off.

h) **The chain of events started when** a terrorist bomb at Gran Canaria International Aiport caused a number of flights to be diverted to Los Rodeos Airport, Tenerife, where it was very foggy.

---

i) **First,** air traffic instructions were standardised internationally.

j) **Immediately**, fire crews rushed to deal with the burning KLM plane.

k) **As a result,** the KLM captain thought that he had received take-off clearance, whereas the controller had not given specific clearance, and started its take-off roll along the runway.

l) **As a result of the accident**, the following measures were taken.

m) **At this point**, the KLM pilot saw the approaching lights of the Pan Am aircraft, tried to take off, but struck the plane with its engines and undercarriage, and crashed in flames.

n) **Third**, ground radar was installed at Los Rodeos Airport (renamed Tenerife North Airport).

o) **At the start of the rescue operation**, fire crews were unaware for 20 minutes that the Pan Am plane was also involved in the accident, because of heavy fog.

p) **Second**, cockpit procedures were changed, with more emphasis being placed on team decision-making.

---

2 Rewrite these statements about the accident, using the third conditional. Replace the italicised words with the words in brackets where applicable.

Example: *1 The crash would not have happened if Gran Canaria Airport had not been closed.*

1 The crash happened; Gran Canaria Airport was closed.

2 The controller was not able to see the aircraft on the ground; he couldn't give clearer instructions.

3 The airport wasn't equipped with ground radar; the controller *couldn't* monitor the aircrafts' positions in poor visibility. (able)

4 Communication between the controller and the pilots was *not good*; the KLM captain thought he had clearance for take-off. (better)

5 The KLM flight engineer did not repeat his concern about the controller's instruction; the pilot *did not* abandon his take-off. (might)

6 The pilots were not able to see each other; the fog was *very* thick. (less)

7 The fire service did not go immediately to both planes; they were not able to see them.

8 The KLM plane took on extra fuel, the blaze *was extremely* severe. (might / less)

3 Communication

1 Read this dialogue, which shows poor communication between a junior technician (A) and the manager of a power station (B). Match the technician's lines (1–7) with the guidelines below that show what he should have done.

A: (1) Oh, dear. Er, uhm, I don't like the look of that.
B: Are you talking to me?
A: (2) Yes, I'm looking at the river.
B: What are you talking about?
A: (3) In my opinion, it looks like there could be a flood.
B: There have never been floods at this substation and I've worked here for 35 years.
A: (4) Shall I do anything?
B: No, if it floods, it floods. I've got work to do.
A: (5) OK, I won't disturb you anymore.
B: Good!
A: (6) I'll go and check the flood warnings online.
B: That's enough worrying! Go back to your work and let me get on with mine.
A: (7) OK, boss, as you say.

Guidelines

a) _____ Try to get your boss's agreement to an action plan.
b) _____ Try more than once when your boss replies with indifference and later opposition.
c) _____ Suggest one or more specific solutions.
d) _____ Be very specific about the problem.
e) _____ Ask your boss to stop their work and think jointly about the problem.
f) __1__ Get your supervisor's attention at the start of the conversation.
g) _____ Give full expression to your concerns.

2 Complete the alternative parts 1–7 below for the dialogue in 1, showing better communication. Choose the most appropriate alternative from the box for each gap.

> duties / concerns   emergency / incident   full / great   possibly / definitely
> river height / flood alert   Sorry / Excuse me   worsening / changing

1 __Excuse me__, sir!
2 We've got an _____. The river's rising very fast.
3 If the river continues rising, there will _____ be a flood.
4 Shall I check the _____ status on the Environment Agency website, boss?
5 Could we give this flood danger our _____ attention?
6 Can I help draw up an action plan and a list of _____ now?
7 The situation's _____ and there's a flood alert for our town. We must discuss our action now.

4 Word list

NOUNS	COMPOUND NOUNS	ADJECTIVES	VERBS
assertiveness	airspeed indicator	accusing	analyse
backup	autopilot system	emotional	buckle
break-up	crisis situation	extensive	contaminate
chamber	fire prevention	faulty	disable
cockpit	high-altitude thunderstorm	investigative	ensure
co-pilot	hydroelectric power station	physical	insulate
concern	in-flight communications	preliminary	shoot
cross-section	intake gate	toxic	signal
dam	metal fatigue	**ADVERBS**	vibrate
debris	pitot tube	assertively	**PHRASAL VERBS**
filtration	radio transmission	excessively	break away
fuselage	service life	offline	put back
housing	short circuit	online	rule out
outcome	take-off speed	**NOUNS (REPORTS)**	step back
penstock	turbine hall	abstract	
plant	use-by-date	attachment	
refund	video footage	background	
respect		conclusion	
reservoir		criticism	
vibration		finding	
warranty			

1 Complete this text, using the correct form of words from the Word list.

(1) ___*Extensive*___ flooding in Hungary resulted from a breach in the (2) _____ wall surrounding a (3) _____ belonging to an alumina (4) _____. Houses in nearby villages were flooded to a depth of 1.5 metres and streets were filled with (5) _____ swept along by the floods. Rivers were also (6) _____ by the (7) _____ sludge. Further consequences of the flood are detailed in the (8) _____ section of the report.

One of the (9) _____ of this report is that heavy rains in the weeks before the disaster led to the collapse of the dam wall. (10) _____ reports also suggest that the soundness of the dam was not adequately monitored and the reservoir had been constructed on an unsuitable site. An (11) _____ of the full report is available online as an (12) _____.

1 ▸ 🎧 29 Write the words in the box on the correct lines, according to the pronunciation of the 'a' vowels underlined. Then listen and check.

> abstr<u>a</u>ct <u>a</u>nalyse att<u>a</u>chment b<u>a</u>ckup d<u>a</u>m fatigue filtr<u>a</u>tion fusel<u>a</u>ge h<u>a</u>ll
> indic<u>a</u>tor int<u>a</u>ke st<u>a</u>tion vibr<u>a</u>te

1 'a' as in 'cap': ___*backup*___ _____ _____ _____ _____
2 'a' as in 'say': _____ _____ _____ _____ _____
3 neither 1 nor 2: _____ _____ _____

1 Projects

1 Complete the evaluation report below of a compressed air energy storage (CAES) system, using the words and phrases in the diagram and the box.

> aim air compressor criteria electricity evaluation generate generator
> grid limestone cavern off-peak electricity peak-period electricity recuperator
> rely requirements turbine

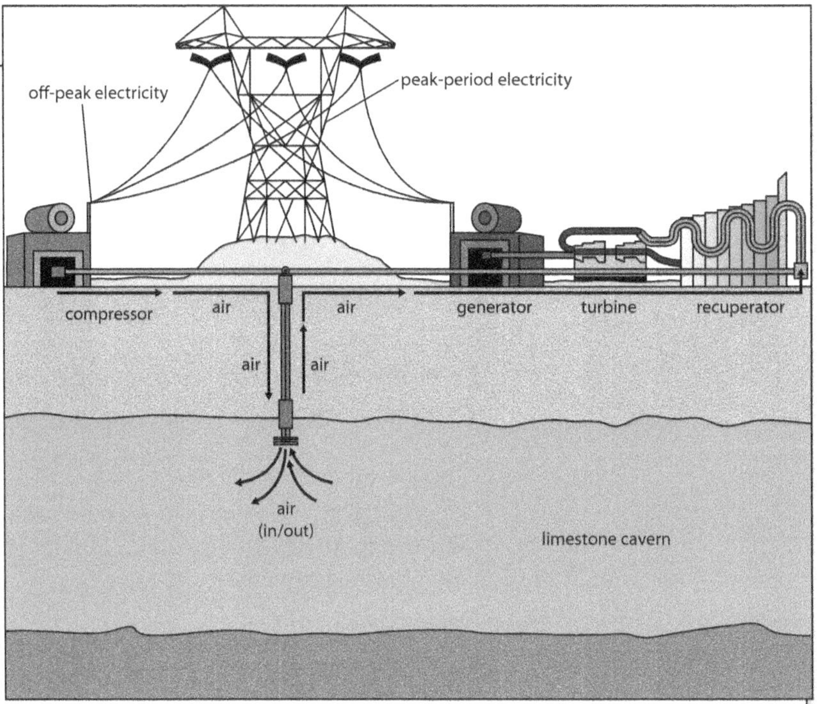

off-peak electricity · peak-period electricity · compressor · air · generator · turbine · recuperator · air · air · air (in/out) · limestone cavern

Introduction This report gives the results of our recent (1) _evaluation_ of the Compressed Air Energy Storage system.

Project objective The overall (2) _____ of the project was to examine alternative energy storage systems for bridging power and energy management of the (3) _____ grid. CAES is one method of storing energy during periods of low demand, releasing the energy and generating electricity at periods of high demand.

Technology CAES consumes less than 40% of the gas used in conventional gas turbines to produce the same amount of electricity. Whereas conventional gas turbines consume about two thirds of their fuel to compress (4) _____ at the time of generation, CAES pre-compresses air using (5) _____ from the electricity (6) _____ at times of low demand. A (7) _____ compresses the air, pumping it into a mine or natural underground reservoir, e.g. a (8) _____ from which oil or natural gas has already been extracted. Later, the compressed air, having been released up the riser and after passing through the (9) _____, is combined with gas fuel to turn a (10) _____ connected to a (11) _____. The electricity thus goes into the grid as (12) _____ when demand is highest.

Criteria Four (13) _____ were used to evaluate the potential of CAES as an energy storage system. These criteria were:

1 Practicality: the system had to be demonstrated in existing installations.
2 Flexible operation: the stored energy had to be able to (14) _____ electricity within 15 minutes.
3 Inputs: it needed easily obtainable inputs, preferably by not having to (15) _____ on fossil fuels.
4 Lead-time: it had to be possible to design and build an installation within 5 years.

Evaluation The project was evaluated as potentially successful against all four (16) _____.

2 Performance

1 ▶ 🔊 30 Listen to an appraisal interview. Are these statements *true* (T) or *false* (F)? Correct the false ones.

1 Some results are incomplete but will be ready next week.
2 The project ran behind schedule and was completed late.
3 It wasn't possible to modify the test schedule once the project was under way.
4 The researcher has learnt four lessons about running a project.

2 ▶ 🔊 30 Listen again and complete the notes in this appraisal form. Some sections are discussed out of order.

Target: test and report on new range of sodium sulphur batteries	
Put Y in box if target has been met in full, otherwise put N	*N*

1 Extent to which target (not) met? *Test completed (but late). Report* _____

2 Reason for any non-achievement? *Delays before and during test programme:*
  2.1 _____ *laboratory space*
  2.2 *unable to obtain* _____
  2.3 _____ *for preparing the necessary* _____
  2.4 *ran* _____ *test*
3 Conclusions from test results? *Tests show* _____ *at most temperature ranges.*
4 Attempts to solve problems? *Modified the test schedule.*
  Attempts successful? *Yes, brought forward* _____ .
5 How to improve next time?
  5.1 *Need to recheck the* _____ .
  5.2 *Should have a* _____ .
6 Lessons learned for starting projects? *Spend more time planning the start-up of a new project.*
  6.1 *Check* _____ .
  6.2 *Check* _____ .
  6.3 *Order* _____ .

3 ▶ 🔊 31 Complete these parts of the conversation in 1 and 2, using the correct form of the phrasal verbs in the box. Then listen and check.

> come up come up with get hold of get on with go about go ahead with
> hang on put off set up turn out

1 Have a seat, Lisa. Do you mind _____*hanging on*_____ a second while I send this email? … Good! That's gone. So, how do you think you _____ the project?
2 How did the test results _____?
3 I think I've _____ some interesting results, which I sent you last week.
4 For example, how did you _____ scheduling the project?
5 Then I couldn't _____ a particular measuring instrument, so I had to _____ the first experiment for a week. And I reckon I underestimated the time needed to _____ the test equipment.
6 Yes, I should have spoken to you as soon as the problems _____ .
7 Yes, but then you _____ one extra test at a different temperature, which there was no need to do.

3 Innovation

1 Complete this text about nanotechnology with the words in the box.

> graphene   graphene sheet   graphite   nanofluid   nanogenerator   nanometres
> nanoparticles   nanoscale   nanoscale pores   nanowires   smart dust

The (1) _nanoscale_ measures dimensions between one and one hundred
(2) _____ (nm). There are currently several nanotechnologies under
development. The concept of (3) _____, a network of tiny micro-sensors,
was presented in 2001. In the future, devices using this concept will be used
mainly for sensing, e.g. for detecting light, temperature, vibration, magnetism or
chemicals. (4) _____ are used for generating electricity from motion. A future
(5) _____ could convert movement into electrical current that can power
mobile devices. In (6) _____ cooling, (7) _____ made of zinc or copper
are added to water to improve thermal conductivity. Engine-cooling systems like pumps
and radiators could be made more efficient, cheaper, lighter and smaller.

In 2010, a Noble Prize for physics was awarded for research into (8) _____,
a material made up of a single layer of carbon. This new two-dimensional
(9) _____ is a near-transparent conductor, just one atom thin. Writing with a
pencil produces thin layers of (10) _____ that are transferred to the sheet of
paper. In medical research, DNA molecules can be detected when they pass through
(11) _____, or microscopic holes in the graphene sheet.

2 🔊 32 Listen to a talk about the properties of graphene. Underline the correct
alternatives and complete the notes in the table. Label the diagram with the correct
measurements.

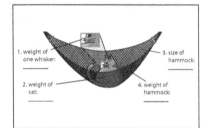

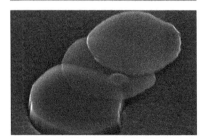

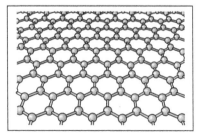

1	Chemical composition	single atomic layer of carbon
2	Structure of graphite	stacked sheets of carbon, with a _____ structure
3	General properties	*thick / thin strong / weak rigid / flexible* *stretchable / malleable* *conductor / resistor* thermal conductivity: *high / low* conductivity: *high / low*
4	Fabrication	now possible to fabricate sheets of *7/17 mm / 17/70 cm* width
5	Current applications	_____ screens, _____ panels, _____ cells
6	Potential applications	_____ electronics, _____ sensors, composite materials for _____ and _____
7	Strength	breaking strength: _____ ; _____ stronger than the strongest steel
8	Density	(see diagram on left)
9	Optical transparency	*coloured / non-coloured*; absorbs _____ of light intensity, *so totally / almost* transparent
10	Conductivity	higher conductivity than _____
11	Thermal conductivity	conducts heat _____

4 Word list

NOUNS	NOUNS	VERBS	PHRASAL VERBS
achievement	objective	assess	come up
appraisal	participation	attempt	come up with
assertiveness	performance	copy (somebody in on something)	get hold of
assessment	schedule		get on (with)
cardboard	scope	expand on	go about
co-operation	seawater	humidify	go ahead (with)
contingency plan	tender	modify	hang on
criteria (plural)	water vapour	reckon	turn out
criterion (singular)	(NANOTECHNOLOGY)	rely on	ADJECTIVES
drop	graphene sheet	roll out	(go) bankrupt
effort	graphite	saturate	essential
electricity grid	nanofluid	trickle	super-saturated
evaluation	nanogenerator		
gravity	nanometre		
greenhouse	nanoparticle		
nutrient	nanoscale		
	nanoscale pore		
	nanowire		
	smart dust		
	transparency		
	transparent		

1   A candidate at a job interview is talking about her present job. Complete the text with nouns from the Word list. There may be more than one possible answer.

What are my (1) __criteria__ for good (2) p _____ at work? Obviously, meeting sales targets, for a start. In my present company, they use a form of management by objectives, so I have sales targets for the coming six months. ... At my last (3) a _____ interview with my boss, I got high scores for co-operation, (4) p _____ and (5) a _____. As a result, I was given a salary increase at my annual salary (6) a _____, so I was pleased with that. ... My best (7) a _____ last year was following up contacts we made at a trade fair. I put a lot of (8) e _____ into that and was able to convert over 60% of enquiries into firm orders. I've also been involved in the (9) e _____ of tenders for goods and services which we buy in. ... I've enjoyed working there, but it's a small company and there's little (10) s _____ for me to advance my career.

2   📻 33 Listen to and repeat the highlighted words and phrases in columns 1 and 2 of the Word List. Underline the syllable with the main stress in each word or phrase.

Section 1

1 Look at the map and read paragraphs A–I below, which describe a sequence of events about a plane crash, but in the wrong order. Mark the points on the aircraft's flight path on the map with paragraph letters A–I. Use the phrases printed in bold to help you.

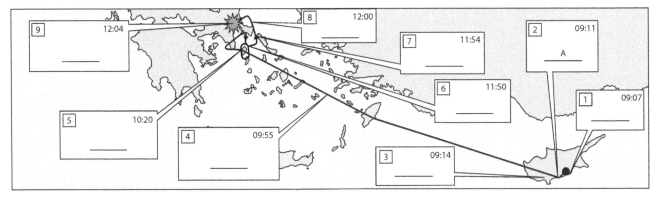

A: Flight Data Recorder (FDR) information from the crash proves that the crew and passengers lost consciousness as the aircraft climbed away from Larnaka airport. It also indicates that the warning horn alerting the crew about the lack of cabin pressure **started sounding continuously four minutes after take-off**.

B: After the flight failed to contact air traffic control **upon entering Greek air space**, two Greek Air Force fighters were despatched to intercept the plane. One of these two pilots saw oxygen masks suspended in the passenger cabin.

C: Helios Airways Flight 522 **took off from Larnaka**, near Nicosia in Cyprus, on a flight to Athens shortly after 9 o'clock in the morning.

D: **About ten minutes before the aircraft crashed** into a hillside, the CVR (Cockpit Voice Recorder) recorded three attempts to send an emergency radio transmission to Athens.

E: **At the moment of impact**, the aircraft's radio was still set to the Nicosia air traffic control frequency, indicating that the pilots had lost consciousness sometime after taking off from Cyprus. 121 people died in the crash.

F: **Three minutes later**, a master caution warning light was activated, indicating that the cabin had failed to pressurise on its ascent. Then at about 14,000 feet the oxygen masks in the cabin automatically deployed.

G: **Ten minutes after the left engine failed**, the right engine flamed out at an altitude of 7,100 feet and from that moment the aircraft was powerless until it crashed.

H: For most of its flight, the aircraft flew on its autopilot system, flying straight over the island of Kea **at 10.20**. It then circled and joined the holding stack for Athens airport, making multiple turns around the Kea holding circuit.

I: **After circling for about an hour and a half**, the left engine flamed out and ceased to provide power, and the plane's descent began.

2 Join these pairs of sentences, using the third conditional, and the words in brackets where applicable.

Example: *1 If the operating company had had good organisation and management, the aircraft would have been maintained in good operating order.*

1 The operating company didn't have good organisation and management. The aircraft was not maintained in good operating order.

2 The airline was allowed to operate unsafe aircraft. The regulatory authority didn't take its safety responsibilities seriously. (more)

3 The manufacturer didn't respond to previous pressurisation incidents in the same type of aircraft. The malfunction occurred.

4 The plane crashed after quite a long time. It was in autopilot mode. (much sooner)

5 The crew didn't share a common language. They weren't able to communicate very well with each other and with the maintenance department. (better)

6 They found the FDR. The causes of the accident were clear. (less)

Section 2

1 Complete the interview below with the manager of a greenhouse complex (B), by doing the following.

1 Write these questions in the correct place in the dialogue.
1 What else do I need to know about greenhouse technology?
2 So, what's your evaluation of your business plan so far?
3 Was it necessary to complete the whole complex at the outset?
4 Hang on – don't tomatoes need soil to grow in?
5 What did you have to do to set up this huge greenhouse complex?
6 How did you go about planning the project?

2 Put the verbs in brackets into the correct form, and fill in the other gaps with the words and phrases in the box.

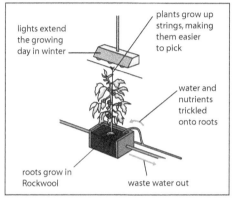

lights extend the growing day in winter

plants grow up strings, making them easier to pick

water and nutrients trickled onto roots

roots grow in Rockwool

waste water out

drops essential gravity greenhouses nutrients
objective produce water vapour

A: (a) *What did you have to do to set up this huge greenhouse complex?*

B: We had to find a suitable site. Fortunately, there (1) *was no need to* (be / no need) to improve or level the site, as it had been used for growing vegetables beforehand.

A: (b) _____

B: No, we (2) _____ (need) plan it all, but we (3) _____ (be / not obliged) to construct all seven greenhouses at once. So, we opened four of the seven greenhouses to start with.

A: (c) _____

B: For one thing, it was (4) _____ to plan our water requirements from the outset. We collect and recycle as much of our own water as possible. Rainwater, having been collected on the greenhouse roofs, flows down by (5) _____ into one of seven large water reservoirs on the site. The water collection system provides about 55% of our water needs. A further 20% is provided by collecting and recycling water from the growing areas. In addition, more water, (6) _____ (evaporate) and become (7) _____, is collected after condensing on the roofs.

A: (d) _____

B: These days, supermarkets rely on growers to provide a regular supply of uniform (8) _____. Our crops are all grown hydroponically, that is they are grown in a fibrous material like Rockwool insulation instead of soil.

A: (e) _____

B: No, they don't. Tomatoes have been grown in soil-free materials for the past 20 years. The baby plants are bought in from specialist growers and placed on the Rockwell growbags. A tube trickles (9) _____ of water and (10) _____ onto each plant, the correct amounts having been carefully calculated by a computer programme, which also controls the lighting. The (11) _____ in the tomato greenhouse is to boost productivity by extending the daylight hours during winter. So the lights are switched on during the night to provide more time for (12) _____ (grow).

A: You make growing vegetables sound very scientific.

B: Which it is, of course. And the software programmes for the control of growing, watering and lighting the greenhouses all (13) _____ (have to / develop / install) before we started producing.

A: (f) _____

B: The project (14) _____ (turn out) very well so far. We now plan to go ahead and construct the three remaining (15) _____ .

Audioscript

Unit 1 Innovations

 02

[H = Hans; A = Ahmed; M = Magda]

H: Hello! I'm Hans, and these are my colleagues, Ahmed and Magda. Our aim in this short talk is, firstly, to explain what an X-ray is; secondly, to describe a simple X-ray machine; thirdly, to explain how it works; and finally, to mention how X-rays are used in medicine.

So, what are the main differences between ordinary light and X-rays? Both of them are the result of the movement of electrons in atoms. In short, the movement of these electrons causes visible light photons and X-ray photons. The difference is that the X-ray photons have much more energy and pass through most things, but not through metals like lead, and not through human bones either.

OK, are there any questions at this point? So, now I'm going to ask Ahmed to talk about the X-ray machine. Ahmed, would you like to take over?

A: Sure. If you have a look at this slide on the screen, you can see a diagram of an X-ray machine. It sits in this lead case. You have a cathode at the top and an anode at the bottom. An electric current passes through the cathode at the top here, and heats it up. The tungsten anode, at the bottom here, draws an electron beam across the tube. This process – I won't go into details now – releases extra energy in the form of X-ray photons.

So what happens to this stream of high-energy photons? Well, first it produces a lot of heat. So there is a cool oil bath here, which absorbs the heat. Next, to stop the X-ray photons from escaping in all directions, we have this thick lead case all around the machine. As a result, the X-ray machine emits X-ray photons in a narrow beam, which passes through a filter, through this small window here, and eventually goes through the patient's body.

When the X-rays pass through the patient, this is recorded on film as a negative image. So the bones appear white and the soft tissues appear black or grey. Why is this? Well, soft tissues are made up of small atoms, so they don't absorb the X-ray photons very well and don't show up on the film. However, the bones are made up of larger atoms, so these absorb the X-rays and show up on the image.

We can now turn to the next part of our presentation, and I'd like to ask Magda to take over. She's going to talk about uses of X-ray technology in medicine. Magda.

M: Thanks. As Ahmed just explained, soft body tissue doesn't show up clearly on an X-ray picture. Therefore, we need to introduce a 'contrast medium' into the body. Contrast media are liquids that absorb X-rays better than the surrounding tissue. For example, to examine the stomach, medics get the patient to swallow a 'barium compound', that is a harmless contrast media mixture. We do this before the X-ray. Then, when the X-ray is taken, the contrast medium shows up and you can see the shape of the soft tissues that you are trying to examine. As you can see in slide A, the patient's stomach is invisible. Now if you look at slide B you can see its shape very clearly and in particular this dangerous swelling here.

That's all I have to say now. Do you have any questions? OK, if there are no questions, I'll hand back to Hans.

03

Ahmed: If you have a look at this slide on the screen, you can see a diagram of an X-ray machine. It sits in this lead case. You have a cathode at the top and an anode at the bottom. An electric current passes through the cathode at the top here, and heats it up. The tungsten anode, at the bottom here, draws an electron beam across the tube. This process – I won't go into details now – releases extra energy in the form of X-ray photons.

So what happens to this stream of high-energy photons? Well, first it produces a lot of heat. So there is a cool oil bath here, which absorbs the heat. Next, to stop the X-ray photons from escaping in all directions, we have this thick lead case all around the machine. As a result, the X-ray machine emits X-ray photons in a narrow beam, which passes through a filter, through this small window here, and eventually goes through the patient's body.

When the X-rays pass through the patient, this is recorded on film as a negative image. So the bones appear white and the soft tissues appear black or grey. Why is this? Well, soft tissues are made up of small atoms, so they don't absorb the X-ray photons very well and don't show up on the film. However, the bones are made up of larger atoms, so these absorb the X-rays and show up on the image.

Unit 2 Design

 04

[C = Craig; A = Anna; B = Basil]

C: Before we go ahead with any plans. I thought we should get together and throw a few ideas about. Who'd like to kick off? Anna?

A: OK. I've been down to the site with our buildings manager. Basically, the old storage building is about a hundred years old. The structure is sound, the floor is OK, but there are problems with the roof and windows.

C: Any thoughts on that? Can we utilise the existing structure and turn it into a new research workshop? What's your opinion?

A: We would need to thoroughly restore the whole building, not just do a bit of redecoration.

C: OK, let's focus on that.

A: We'd need new doors and windows. We'd need to insulate and lay a new floor over the old concrete base. And the roof would have to be replaced.

C: What do you think, Basil?

B: It's possible. We could renovate it to a high specification, and this would be cheaper than demolishing and rebuilding. I suggest that we get an architect to draw up plans for a modernised, totally refurbished building, not a new one. I think that's the most cost-effective solution.

C: Right, that makes sense. Now, we do have a company policy of energy conservation. So we need to brief our architect accordingly. Any more ideas on that?

B: Double glazing for the windows and any external doors. Our architect will specify under-floor insulation. If the cavity walls aren't already insulated – which I expect they aren't – I suggest we insulate those, too. That way we can cut down on our heating bills considerably.

C: That's very interesting.

A: I've got an idea, Craig. What about putting solar panels on the new roof?

C: Brilliant! But what's the orientation of the existing roof?

A: It's south facing, which is perfect. The new solar panels should pay for themselves within ten years, but we'd need to check the figures to find out if it's cost effective. And with the present government scheme, we can sell any electricity we don't use back to the grid.

C: That's a great idea! We can utilise all the south-facing roof area and that will enable us to keep our electricity costs down. Why don't we fix an appointment with a solar panel installer for a site visit? Then we can get some costs before we go any further. Agreed?

A & B: Yes / OK.

05

In the next part of my presentation, I'll go through the properties of the four different materials that will be used for the roof.

Starting with the solar panels, these are exceptionally durable, and in fact they're guaranteed for 20 years. Furthermore, they are self-cleaning and therefore maintenance-free. The other thing is that they provide an additional degree of thermal protection.

I'd like to mention that I'm recommending the cheaper specification, the 125-watt panel, since we can utilise the full roof area. A panel with a maximum output of 125 watts has a module conversion efficiency of 13.3%. It's cheaper to install more panels with a lower specification, instead of fewer panels with a higher specification.

Let's move on to the roof tiles. I've recommended artificial tiles, since availability is good, and they have a good strength-to-weight ratio and are long-lasting, as you know. Most artificial tiles will last for a hundred years, though of course this is not guaranteed.

The next layer, working from the outside to the inside, is the breather membrane. Here I'm specifying a general insulating breather membrane with good thermal protection. Its tensile strength is sufficient, since it rests on a level surface and will not be permanently stretched. The tensile strength is the same both *along* the material and *across* the material, so it's easy to handle. It's flexible and easy to cut. It's also very thin, 0.25 millimetres, so it's very light. The weight, by the way, is 85 grams per square metre.

Moving on now to the insulating material; I'm specifying a high-performance thermal insulation board. This is ideal for using between and below the rafters. We'll use a 45 millimetre thick board between the rafters, and a 25 millimetre thick board below the rafters. This will provide excellent thermal protection and long-term energy savings. So we'll finish up with an agreeable ambient working temperature for everybody. Also, it minimises the additional loading on the load-bearing walls, as it's so light: the weight of the thicker board is 1.4 kilograms per square metre, and for the thinner board, the weight is 0.85 kilograms per square metre. Installation is easy, as it's easy to cut to shape. Are there any questions you would like to ask? OK, well in that case …

See Answer key page 71, Word list exercise 3.

Unit 3 Systems

[N = Newsreader; R = Reporter]

N: And now we have news of another product recall from the Murray car company. Here's our Industrial Correspondent Bo Hagen.

R: Murray, the car company which had several product recalls for 2009 models, has just announced a further recall from their production line of 2010. The recall concerns the Barrossa model, and is limited to new car registrations for the year 2010. A statement from the company 'urges all customers who bought any of the Barrossa range of vehicles to contact their local dealer without delay. The fault is with the power steering gear assembly, and some low-speed accidents have been reported.' That's the end of the statement. So, the problem concerns the power steering system, and the company's advice is to get it checked out as soon as possible. Further news from the car industry: half-yearly results are expected this morning from …

08

[I = Interviewer; S = Spokesperson]

I: Would you mind answering a few questions? I'm sure that a lot of your customers will be very concerned to hear about this product recall.

S: Of course. They should take their cars back to their local dealer as soon as possible to have the problem rectified.

I: I understand that the problem is with the power steering. Now, that's kind of critical, isn't it?

S: Yes. There's a problem with the lower pinion bearing, which forms part of the power steering gear assembly.

I: Sorry? What is it?

S: It's the lower pinion bearing. And that is part of the power steering gear assembly that turns the front wheels.

I: So what is the worst-case scenario?

S: The lower pinion bearing could, in some cases, separate.

I: I see, and would that be serious?

S: Well, yes, it could actually permit movement of the pinion shaft.

I: I see. Have there been any serious accidents?

S: None so far and no fatalities, I'm happy to say. But we have received reports to date of 11 accidents with related injuries, so we need to get this problem sorted as quickly as possible.

I: And what should owners look out for? What kind of malfunction might they notice?

S: Well, drivers have reported an intermittent loss of power steering when making left turns, usually at low speeds.

I: And is it easy for dealers to rectify this problem?

S: Yes, it's very simple. Either the dealer will install a new lower pinion bearing. Or else they will replace the whole steering gear assembly.

I: And how much will this cost? It sounds expensive.

S: The repairs will be done free of charge. So there's no cost to the customer, either for parts or for labour.

See Answer key page 72, Word list exercise 3.

Unit 4 Procedures

Hello. Welcome to Diamond. In fact, the full name is Diamond Light Source Ltd, but we all call it Diamond for short.

Diamond is one of about 40 large synchrotrons in the world. Now what exactly is a synchrotron, you may ask? A synchrotron is a huge circular ring which produces intense light with the help of electrons. Diamond is a particle accelerator, that accelerates electron beams almost up to the speed of light.

Now, many people ask if Diamond is the little sister of the Large Hadron Collider, the LHC, at Cern. The important point is, they are both particle accelerators, but different types. The difference is in the names. Cern is a collider, for studying high-energy particle collisions. Diamond is a particle accelerator which generates intense synchrotron light to study a range of samples at the atomic and molecular level.

Just a few facts for you: work on the construction of Diamond started in 2003 and was completed in 2007, and a lot of research has already been done here. About 400 people work here, and research using X-rays, ultra-violet and infrared light has been carried out in different areas, including diseases like cancer, Alzheimer's and malaria, potential new drugs, and examining metal structures for signs of metal stress.

11

Let's now have a look at this 3D model of Diamond. One of the problems with Diamond is that much of the activity is happening at the particle level, and of course, particles are invisible. Also the light we are generating is 100 billion times brighter than the sun. So we can't show you that either!

Do you see this machine at the centre here, at the end of a long straight tube? This is where it all starts. That is an electron gun which generates particles called electrons. Some of you may remember the cathode ray tubes in old TV sets. It's the same principle. The electrons travel down this straight tube here called the linac which is short for linear accelerator. The electrons then enter this circular ring in the middle, which is called the booster synchrotron. Here they accelerate up to even faster speeds. Then they leave the booster synchrotron and move into the storage ring through this tube here.

Let me explain about the storage ring. It looks like a circle, but in fact it's made up of straight lengths of tube which are angled together with bending magnets. These bending magnets steer the electrons around the storage ring and stop them crashing into the sides of the walls.

Have a look at these angled tubes coming off the storage ring, called beamlines. There are currently 19 of them in operation, and we can increase that number in future. Now, remember that the electrons are travelling clockwise around the ring, and the beamlines are for light that is *leaving* the ring. Light travels out of the storage ring and into the beamlines, which are experimental research stations where the measuring instruments are located.

12

[G = Guide; V1 = Visitor 1; V2 = Visitor 2]

V1:   How is an electron beam produced?

G: Electrons are produced in the electron gun. A high-voltage cathode is heated up under a vacuum. This allows the electrons to escape in a stream.

V1:   What are the dimensions of the electron beam?

G: On average, the beam is about a quarter of a millimetre wide, by about a sixtieth of a millimetre in height. Any other questions?

V2:   Why must the electron beam be contained in an ultra-high vacuum?

G: If the electrons travelled round the storage ring through air, they would collide with air molecules. To avoid this, they travel in a vacuum that is one trillion times lower than atmospheric pressure. Another question?

V1:   How long do electrons remain in the storage ring?

G: Electrons remain in the storage ring for around 20 hours. As they travel, they collide with a very few particles that are still in the vacuum and are lost. New electrons are added to the ring twice a day. In future, this topping-up process will continue without a break.

V1:   What kind of electron beam is the best?

G: Ideally, we need an electron beam that is stable and that produces light with higher average brightness. Is there anything else you'd like to know?

V2:   What causes synchrotron light?

G: Synchrotron light is emitted when a beam of electrons moves at close to the speed of light and is bent by a powerful magnetic field.

Unit 5 Processes

 13

On behalf of our company, welcome to a presentation about our new FINEX Process Technology. The location of our FINEX steelmaking plant is in Pohang, in South Korea. The plant was inaugurated on May 30th 2007, at the end of a three-year construction period, which began in August 2004. Initially, the output was 1.5 million tonnes per annum.

The next major project for FINEX Process Technology has already been given the go-ahead. The steelmaking plant is being built in the State of Orissa in India. Phase 1 was completed in 2010, with an initial capacity of four million tonnes per annum. The plant is planned to have a final capacity of twelve million tonnes per annum.

To view a specification chart for the steelmaking plant at Pohang, select 1. To watch an interview with the CEO, select 2.

 14

[I = Interviewer; C = CEO]
I: Now that the new plant is up and running, what would you say are the main advantages of the FINEX Process Technology which your company has developed?
C: It's an eco-friendly iron making process and it allows us to use cheap iron ore powder and non-coking coal as feedstocks. In addition to lower operating costs and emissions than the traditional blast furnace method, we have reduced our construction costs by eliminating the need for coking and sintering plants.
I: Could you explain why construction costs are lower?
C: Of course. Basic oxygen iron making needs two plants that the FINEX Process doesn't require. First, the sintering plant. The process of sintering refers to the preparation of iron ores for the blast furnace. To be specific, sintering plants produce iron ore pellets. We don't need a sintering plant because we use a different type of iron ore feedstock. The FINEX Process uses a powdered iron ore, which has the advantage of being about 23% cheaper.
I: I see. You also mentioned coking plants.
C: Yes. Basic oxygen iron making requires a coking plant, which produces coke from high-class soft coal. FINEX doesn't require a coking plant, as we use general coal, which is widely available and is also 20% cheaper.
I: I see. So your plant construction costs are lower because you don't require the sintering and coking plants.
C: Correct.
I: And your production costs are also lower because you are buying in cheaper feedstocks, both the coal and the iron ore that you use.
C: Correct. And there's one more point to make about the iron ores that we shall be using in our new plant in India. We shall be buying in Indian ore, which has the advantage of a high-aluminium content.
I: Sorry?
C: The Indian iron ore has a high content of aluminium, which suits the FINEX Process even better. In total, our production costs are between 10 and 15% lower with FINEX.
I: Going back to the subject of pollutants, you stated that the emissions in a FINEX plant are lower. Could you be more specific about that?
C: Yes. Emissions of sulphur and nitrogen oxide are significantly lower in a FINEX plant, as well as emissions of carbon dioxide.
I: How much lower? What percentage?
C: I don't have a figure to hand, but the emissions of all three pollutants are higher in the BOS method, mainly because of the sintering plants that they have. FINEX produces fewer pollutants, and so we have lower costs in eliminating them.
I: You mentioned lower production costs and also lower operating costs for FINEX. Could you explain those a little bit more?
C: Yes. As we don't have sintering or coking plants, we need fewer maintenance staff and we also need less production time. So we save money, and this is a big factor in our company remaining competitive in future.
I: Thank you very much.

 15

See Answer key page 74, Word list exercise 2.

Unit 6 Planning

 16

[I = Interviewer; C = CEO]
I: Now, you're planning to construct a new nuclear power plant at Medupi. Could you tell me more about that, and your country's nuclear power plant programme?

C: Of course. Let's start with the plant at Medupi. This plant will have a power output of 4,800 megawatts. The budgeted construction cost is 125 billion rand, which is the equivalent of 17.5 billion US dollars. Moving on to our NPP programme, we plan to construct a total of five or six reactors, and we hope to create 70,000 jobs.
I: Will it be easy to get this huge project moving?
C: We have identified some major future needs. First, we shall need to train more experts. Second, we shall need to build up a whole nuclear energy support industry, for example, for the construction of the reactors, for the maintenance of the plants once they're operating and of course to deal with the nuclear waste.
I: Isn't it incredibly expensive to build a nuclear power plant?
C: Yes, in fact it's about three times more expensive than building a coal-fired power station. But there are economies. It will be cheaper to build a series of NPPs than to build a single one. So there are savings.
I: What about the fuel? Is it difficult to get hold of the uranium that a reactor needs?
C: Yes and no. First of all, South Africa already mines uranium, so we have locally produced uranium already available. However, we aren't yet able to process it. We have plans to convert uranium in the Republic of South Africa for use in nuclear power plants. At the moment, we have to export the uranium and then re-import the final product, the uranium reactor fuel.
I: Why isn't South Africa pursuing a policy of renewable energy? After all, there's plenty of sunshine and wind, for a start.
C: We are doing that in parallel. This year we expect to generate an extra 1,025 megawatts through our renewable energy programme. And that figure is set to rise.
I: You mentioned the higher construction costs of building NPPs. So what are the benefits?
C: Right. In spite of the higher construction costs, and higher operating and maintenance costs, we have much lower fuel costs, compared with a coal-fired plant. So overall, coal and nuclear plants have the same total operating costs.
I: There were a lot of scares about nuclear power plants a few years ago. So is the nuclear industry at present in decline, or growing?
C: Worldwide there are already 440 nuclear power plants operating in 30 countries. And there are as many as 60 reactors currently under construction in 15 countries. So there is a definite upward trend for nuclear power generation as we speak.
I: And which countries are currently investing in NPP programmes?
C: Most of the new plants are being built in Asia, and there are further ones planned for Europe, Russia and the USA. However, the increase is not just due to the building of new plants. In addition, further nuclear capacity is being created by upgrading existing plants and also by extending their working life. We can now keep a reactor running for much longer than we once expected.
I: Thank you very much indeed.

 17

Let's talk about some of the obvious benefits of the new Medupi nuclear power plant. First of all, I've seen the plans for the power plant, and it is aesthetically stunning. Next, it's very well located on that particular stretch of the coast: it's close to the economic powerhouse of Medupi, and it's also close to an important manufacturing hub.

Another good thing is that they haven't used prime agricultural land. Instead, it's part of a land reclamation programme. They plan to construct an artificial island for the reactor installation. So I'm upbeat about the benefits.

However, what about the risk assessment of the NPP programme and what does it disclose? Naturally, there are potential risks of radiation, and the consequences of any radiation leaks would be catastrophic for the local population. The risk assessment also mentions the serious possibility of seawater contamination. As we know, seawater contamination affects people who live along the coast, as well as the safety of fish and seafood over a very wide area of ocean.

But what, you may ask, is the most important environmental concern about the new NPP? The biggest concern is the imminent rise in global sea levels. A nuclear power plant which is located next to the sea and which is dependent on seawater for cooling could be flooded in future. So we need to check that it can continue to work during its operating life.

Unit 7 Developments

Welcome to this introduction to the subject of fibre optics. Fibre optics are an essential part of telecommunications networks. Basically, they are long thin tubes of very pure glass, arranged in bundles called optical cables. The purpose of fibre optics is to transmit light signals over long distances. Now, I've got a slide here of a fibre optic line, or 'fibre', as it's sometimes called. In the centre you have the core, that is the hollow glass tube down which the light travels. The cladding – here – is the reflective material surrounding the core and this reflects the light back into the core. On the outside – here – you have a buffer coating made of plastic, which protects the fibre optic line. Fibres are very narrow; they can be from 9 microns to 62 microns in diameter. To put this in everyday language, they have the same diameter as a human hair. Transmitters send pulses of non-visible infrared light along the core. What kind of transmitters do we use? Well, laser transmitters are more powerful sources of light than LEDs, but they vary more with ambient changes, or to put that another way, they are more variable in their performance, depending on factors such as temperature and humidity.

Look at the second slide. Here you can see that the light bounces from the cladding that lines the core and this is called 'total internal reflection'. Think of putting up a lot of mirrors along the sides of a very long corridor. Because the cladding doesn't absorb any light from the core, the light wave can travel great distances. Another way of putting it is that this system allows signals to go round bends and travel a long way. In due course, the light signal degrades, that is to say, it becomes weaker and less pure. This depends on the distance, of course, but also on the purity of the glass.

Now, what are the advantages of fibre optics? Previous telecommunications systems used copper wires. These had many disadvantages, compared with fibre optics. For a start, the copper lines were very expensive, quite thick and also heavy. You needed more relay stations than you need for fibre optic systems. In addition, signals through copper wires suffered greater degradation, that is, there was often a loss of signal, and voices on the telephone appeared faint. Or else the signal from another line interfered with your signal. This meant that you could often hear another conversation while you were talking. Now, unlike copper wires, fibre optics do not suffer from interference from other fibres in the same cable.

Now I'll hand over to Mona, and she's going to talk about fibre optic relay stations. Mona, over to you …

See Answer key page 76, Word list exercise 2.

Unit 8 Incidents

On 6th July, a tanker driver in southwest England arrived at the gates of the Lowermoor Water Treatment Works. There were no employees at the plant, but the driver had a key for one of the tanks there. After unlocking the cover, he unloaded into the tank 20 tonnes of aluminium sulphate $(Al_2(SO_4)_3)$, which is used to help collect impurities as part of the water treatment process.

However, the tank into which the aluminium sulphate was tipped contained drinking water for the local population. The water supply to a local population of 20,000 people plus 10,000 tourists was polluted. The maximum recorded aluminium concentration in the drinking water ended up as 620,000 micrograms per litre, compared with the maximum concentration permitted at the time of 200 micrograms per litre.

[S1 = Student 1; L = Lecturer; S2 = Student 2; S3 = Student 3]
S1: Was the water supply badly contaminated as a result of the incident?
L: Without a doubt. Over 60,000 fish were killed in two rivers following this incident. They kept pumping water through the network for months and this stirred up some of the soft mud in the pipes.
S2: Did the water authority warn consumers that the water was polluted and unsafe to drink?
L: On the contrary. For several days, the water authority insisted that the water was safe to drink. They even suggested that customers should mix the water with orange juice to mask the unpleasant taste.
S3: In your opinion, was the water safe to drink?
L: Definitely not. What actually happened was this: as the acidic water flowed through the supply network and into people's homes, it corroded the whole network. The result was a highly contaminated water supply with high levels of copper, zinc and lead in it. Definitely toxic and unsafe to drink.

S1: I understand that the water authority did eventually warn customers of the contamination and advised them to boil their water before drinking it. Wasn't this sound advice?
L: Actually, boiling the water made it more toxic, because by boiling it, people concentrated the contaminants and the heavy metals present in the water.
S2: Didn't the company have a programme of routine maintenance at their water treatment works?
L: In fact they did. The water tank that received the aluminium sulphate should have been cleaned out every six months. In fact, it hadn't been cleaned for over three years, and this resulted in a build-up of soft mud, which then went through the water supply network to customers.
S3: In your opinion, was there a problem with management at that particular water treatment works?
L: To some extent, yes there was. Routine cleaning of tanks wasn't happening, as I said before. There were inadequate health and safety systems; for example, tanks weren't clearly labelled.

See Answer key page 77, Word list exercise 2.

Unit 9 Agreements

[PE = Port executive; SM = Salesman]
PE: This is our Security Operations Centre. Have a look at this wall map. This shows the extent of the port facilities and the area that we need to control for security purposes.
SM: Could you tell me about your main security concerns?
PE: Yes. First, we need to keep an eye on all vehicle movements into and out of the port, as well as vehicles moving around inside it. We've had some thefts from containers over the past three months. Incidentally, when incoming vehicles make their way through the port, they often stop and ask for directions. We're in the process of updating all our direction signs.
SM: So how do you control all the security aspects now? What's your present system?
PE: We have a shift system of security patrols on foot. We need to cut down on the salary bill for these patrols. We have a very large area to patrol, as well as a long perimeter fence. One of their priorities at present is looking out for intruders.
SM: For a start, I propose that we secure the perimeter fence with a combination of lights and CCTV cameras.
PE: How should we go about installing a lighting and CCTV network? Won't the cost of putting in extra cabling for electricity be prohibitive, for a start?
SM: We'll look into that, of course, and draw up some figures.
PE: The CCTV cameras must be able to home in on areas inside and outside the port facilities. We've had a series of break-ins where intruders have cut through the perimeter fence.
SM: We'd go for a system of remote control cameras and movement-sensitive lights. A system is as strong as its weakest link, so we shouldn't leave any parts of the port unprotected. That's why you're losing goods right now.
PE: Exactly. We need 100% coverage of the whole port area. If we don't achieve this, it will become obvious which areas are covered by security and which areas aren't. I think we'll go for a total security package, including radio-enabled sensors.
SM: One other question. Do you want a scalable system, or will the port stay the same for the foreseeable future?
PE: Good question. Our nearest competitor is a rather old-fashioned port. It hasn't modernised, and it's running down its operations, so we are in a good position to pick up new business. In fact, we're already in discussions with prospective customers who are thinking of switching their business to us. For that reason it's important that our new security system is scalable, so that in the event of expansion, we'll be able to extend our security coverage accordingly.
SM: Agreed.
PE: So, as a first step, could you draw up a detailed proposal for the systems we've discussed? As soon as you've submitted your proposal, we'll discuss it in detail and get back to you.

[AE = Airport executive; CM = Contracts manager]
AE: Hi, Max. I had a good meeting with the contractor about the new terminal building. I've got a copy of their contract here, and there are a few changes. First, work on site must commence on the agreed date – otherwise the contract will be cancelled.

CM: Sounds OK to me. If they start late, they'll never catch up.

AE: Next, penalty clauses. They need to understand that if they're late, they'll lose money. So, in Section 23.5, we need to add: In the event of bad weather, the penalty clauses for late completion will not be changed. And in Section 23.9, add: Even if the delivery of building materials is delayed, the penalty clauses for phased completion will stand.

CM: Anything else?

AE: Payment. We've agreed on the standard system for payment but there are a few amendments. Section 78.5 is slightly different: Payment will be made monthly for the work certified as completed, on condition that we are satisfied with the work. Section 78.6 is: The contract price is due on completion, with the proviso that 2% is withheld for a period of 12 months, pending making good any defects.

CM: So not the 1% as we've had before now.

AE: No, for such a big project we need to withhold 2%. Now, our contractor insists on the following clause: If the client fails to make the scheduled monthly payments within 14 days, as stipulated under 'Business Terms', the contractor is entitled to withdraw their labour from the site. They say that they can't carry on working if they're not being paid on time.

CM: I guess that's reasonable.

AE: Two clauses here about quality control. Do you see?

CM: Surveyors, building inspectors and the client can inspect the Works as long as notice is given the previous day. OK, that's almost standard. And the second one: In case of disagreement over quality standards, an independent surveyor can be asked to assess the work, subject to approval by both parties. I think we can agree to that. We had something similar on another contract.

AE: One more minor detail – flooding. They might have to pump out water if there's a lot of rain. They call it 'de-watering'. So: De-watering of the site is allowed during heavy rain, provided that the surveyors are informed at the start and end of such activities. And then the costs of the de-watering have to be met by us. And that's all!

See Answer key page 78, Word list exercise 2.

Unit 10 Testing

[E = Eric; C = Carmen; M = Mario]

E: All right. Let's start the meeting now. Right. The first item on our agenda is the Douro Bridge. The client needs us to carry out some non-destructive tests on the steel girders on the bridge and we have to get back to the client with costs and a schedule. Carmen, could you give us some input here? What did you find out about this subject?

C: You have a note of the anticipated problems in front of you. In particular, there may be several affected girders; we don't yet know for certain. Two of the testing methods we were able to rule out immediately as not suitable. One was the eddy current method, because the flaws in the girders may be too deep to detect. Similarly, the magnetic particle method has this problem, too, that is the flaws may be deep in the girders and we can't detect them. In addition, magnetic particle testing doesn't work well with painted surfaces, and of course all the girders on the bridge have been painted.
That leaves us with two possible methods, radiographic testing and ultrasonics. Personally, I favour the ultrasonic method, because it penetrates deep into the girder. Generally, X-rays work well on girders because they can penetrate deep into the girder. However, on this bridge there's a problem of access. We may not be able to get to all the top and bottom surfaces of all the girders. And of course, for X-rays to work, we need to photograph from one side and place the photographic film on the other side. So X-rays could be problematic. And of course this method is more expensive.

E: Thank you, Carmen. That was extremely useful input. Mario, would you like to add anything to that?

M: Is it possible to use *both* ultrasonics *and* X-rays on different parts of the bridge?

C: Yes, you could use ultrasonics for those girders which you can't X-ray, but why have separate equipment and separate operators? That would come out even more expensive.

E: Thank you for raising that possibility, Mario. Before we come to a decision, I'll briefly summarise what we've discussed so far. The eddy current method and the magnetic particle method are both out of the

question. X-rays could be problematic. If ultrasonics are easy, and less expensive than X-rays, why not go for ultrasonics? It's time for us to take a decision. Carmen? Mario?

C: Ultrasonics. Cheaper and no access problems.

M: Ultrasonics. Same reasons.

E: And I can go along with that choice of method. Then that's agreed. Let's move on to the next item on our agenda. Which is … the high-rise apartment block at Faro. Mario, this is your project. What can you tell us?

M: You've got a project brief and some diagrams in your folders. Here, the client needs us to check some steel beams in the apartment block for surface and near-surface flaws. Again, we've got a problem with access, because we may not be able to get close to the steel beams to carry out a visual inspection. I started off by ruling out the ultrasonic method as unsuitable. The problem here is that the linear flaws may be parallel to the sound waves, in which case they may not cause enough reflection.
I considered the magnetic particle method as a possibility. However, with this method, you need visual inspection at the site of the flaws and this may not be possible, as the beams are embedded in the structure. So I've had to rule out this method, too.
X-rays are a good possibility, but they're not the best solution because the flaws may be too thin and they could run parallel to the X-rays. In that case, they wouldn't show up on the film.
Fortunately, we haven't got any problems with eddy current testing, which is therefore my first choice. It's suitable because the flaws are close to the surface and at right angles to the direction of the coil winding. So I suggest eddy current testing.

E: Thank you, Mario. That was a very interesting presentation. Carmen, would you like to comment at this point? Shall we throw this item open for discussion?

C: I agree with Mario. Looking at the architect's plans for the high-rise block, I think access would be a problem for magnetic particle testing. And ultrasonics is out of the question. So, I agree we should use eddy current testing.

E: And I go along with your opinions, too. OK, we need to decide on some action. But I'd like to do that in a separate meeting with the rest of the team. OK?

C: Yes.

M: Agreed.

E: Then that concludes our meeting. Excellent! Thank you both very much.

1 Let's start the meeting now. Right. The first item on our agenda is the Douro Bridge.
2 Carmen, could you give us some input here? What did you find out about this subject?
3 Thank you, Carmen. That was extremely useful input. Mario, would you like to add anything to that?
4 Before we come to a decision, I'll briefly summarise what we've discussed so far.
5 It's time for us to take a decision.
6 Let's move on to the next item on our agenda.
7 That was a very interesting presentation. Carmen, would you like to comment at this point?
8 Shall we throw this item open for discussion?
9 OK, we need to decide on some action.
10 Then that concludes our meeting. Excellent!

Unit 11 Accidents

28

On 4th October 2010, a reservoir dam burst near the village of Kolontar in Hungary, releasing a flood of toxic red mud, or industrial sludge. Nine people were killed in the disaster and around 120 people were injured, some of them seriously.

The sludge was being stored in a reservoir belonging to an alumina plant. Between 600,000 and 700,000 cubic metres of the sludge burst through a breach in the dam wall and flooded an area of 40 square kilometres, affecting at least seven villages and towns, about 160 km from the capital Budapest. The spillage also contaminated rivers and farmland. About 700 people had to leave their homes.

The reservoir at the alumina plant, which employs 1,100 people, covers an area of 10 hectares. Estimates of the amount of red sludge stored at the plant vary between 20 million and 30 million tonnes.

Since the breach in the dam and the flooding of the area, workers have hurried to build another dam, a protective earth wall 620 metres long, with a height of about three metres. This has been constructed between the dam and Kolontar to protect the village from further flooding.

Local people have been asking why so much toxic chemical sludge was held at the reservoir, so close to towns and villages. Normally, this waste, a by-product of the alumina production process, is released into a holding reservoir, where the water evaporates and the waste dries out and turns into hard mud that can be used in the construction industry. In practice, that has not been happening in Hungary. When one enormous reservoir becomes full, another one is built. So there has been no recycling of the dried red mud, although some unsuccessful attempts were made in the past.

The collapse of the dam wall and the cause of the flood is still a subject of speculation. However, heavy rains over the previous weeks had increased the capacity of the reservoir, and the dam wall was insufficiently strong to hold back the toxic sludge. A surveyor has also come forward with another explanation. He had surveyed the site before the reservoir was built 30 years ago and had reported that the unstable soil on the site made the location unsuitable for a reservoir.

See Answer key page 80, Word list exercise 2.

Unit 12 Evaluation

[M = Manager; L = Lisa]

M: Have a seat, Lisa. Do you mind hanging on a second while I send this email? … Good! That's gone. So, how do you think you got on with the project?

L: I had a few problems along the way which I'll explain. But I completed the project successfully. I'm sorry the report was late, by the way.

M: How did the test results turn out?

L: Much better than expected. I think I've come up with some interesting results, which I sent you last week. The details are in the Findings and Conclusions of the report and all the tables are in the Appendices.

M: I note that the new range of batteries has increased efficiency at most temperature ranges. Do we need to run any more tests, or have we finished?

L: We've finished … until the next range of batteries comes along.

M: I note that the testing ran behind schedule, too. Mmm. There were some problems with starting up the project, as I understand. Would you like to expand on the reasons for the non-achievement of your objectives? For example, how did you go about scheduling the project?

L: For a start, I was slow in booking up lab space. Then I couldn't get hold of a particular measuring instrument, so I had to put off the first experiment for a week. And I reckon I underestimated the time needed to set up the test equipment. So I lost a bit of time before and during the test programme.

M: Do you realise that I knew nothing about this at the time?

L: Yes, I should have spoken to you as soon as the problems came up. And I'll know in future that I need to spend more time planning the start-up of a project.

M: OK, I hear that. How did you go about choosing the best order for the tests?

L: I had to modify the test schedule, because of the delays that I've mentioned. I know that I copied you in on the revised schedule. Fortunately, I was able to bring forward two tests which didn't need the missing instrument.

M: Yes, but then you went ahead with one extra test at a different temperature, which there was no need to do. Why?

L: The project was behind schedule, and I forgot to go back to the list of test specifications and check before starting the final test.

M: I'm afraid that counts as another non-achievement on your appraisal form. All right, so how would you avoid that in future?

L: It's a case of 'more haste, less speed'. In other words, I need to recheck the test specification before starting each test to see what's required. And I should also have a contingency plan in case I have to reschedule any tests.

M: OK. What lessons have you learned about starting projects in future?

L: Well, there are three things I need to do to minimise the risk of delays. One is to check the laboratory schedules at the outset of a project. And another is to check the list of equipment needed as early as possible. And of course the third thing to do is order any extra items of equipment.

1   Have a seat, Lisa. Do you mind hanging on a second while I send this email? … Good! That's gone. So, how do you think you got on with the project?

2   How did the test results turn out?

3   I think I've come up with some interesting results, which I sent you last week.

4   For example, how did you go about scheduling the project?

5   Then I couldn't get hold of a particular measuring instrument, so I had to put off the first experiment for a week. And I reckon I underestimated the time needed to set up the test equipment.

6   Yes, I should have spoken to you as soon as the problems came up.

7   Yes, but then you went ahead with one extra test at a different temperature, which there was no need to do.

Today I'm going to talk about graphene. Now, graphene consists of a single atomic layer of carbon. As you know, carbon is the base for DNA and all life on earth, and can exist in several different forms. The most common form of carbon is graphite, which consists of stacked sheets of carbon with a hexagonal (six-sided) structure.

So, how do you make graphene? Have a look at this slide of a piece of graphite. You can see that they've sliced off two flakes, or slices, of graphite from the block, and there they are on the right-hand side. These are two ten-nanometre flakes; each one is 30 carbon atoms thick. Or perhaps I should say 30 carbon atoms *thin*.

Now look at this next slide: here you can see how graphene is made up of an atomic scale network of carbon atoms. And it clearly shows the six-sided structure of each carbon atom. When we talk later about transparency, you can understand why.

The new material graphene has a number of interesting mechanical and electrical properties. It's naturally very thin. It's a transparent conductor. It's much stronger than steel and is very stretchable and flexible. The thermal and electrical conductivity are both very high and it can be used as a flexible conductor. With new industrial fabrication methods, it's possible to fabricate sheets 70 cm wide.

Let's come on now to some applications. Since graphene is a transparent conductor, it can be used in touch screens, light panels and solar cells. Among several possible applications for the future, we can include flexible electronics and gas sensors. In the future, new types of composite materials based on graphene could also be used in satellites and aircraft, mainly because of their great strength and low weight.

I'll now give you some details about its specific properties. First, strength. Graphene has a breaking strength of 42 newtons per metre. Got that? 42 newtons per metre. Graphene is more than a hundred times stronger than the strongest steel. Have a look at this slide of a happy cat in a hammock, or very fine net, tied between two trees. You could place a cat weighing about four kilos in this one-square-metre hammock before it would break.

This brings us to the density of graphene. Go back to our very fine net, or hammock, measuring one square metre and made from graphene. It would weigh 0.77 milligrams. Milligrams, not grams. The hammock – it's so fine it's almost invisible – would weigh less than one milligram, which is the same weight as one of the cat's whiskers. Makes you think!

Next, optical transparency. Graphene is almost transparent, as we saw in the earlier slide, and absorbs only 2.3% of the light intensity. It's colourless. Conductivity. Graphene has a higher conductivity than copper.

Thermal conductivity. Graphene conducts heat ten times better than copper. I see we've got a little time left before the end, so are there any questions?

See Answer key page 80, Word list exercise 2.

Answer key

1 Innovations

1 Eureka!

1
1. 1 was working
   2 was developing
   3 discovered
   4 required
   5 could be separated
   6 launched
   7 was selling

2. 1 was doing
   2 was working
   3 discovered
   4 was invented
   5 to be used
   6 was reduced
   7 were shrunk
   8 could be implanted

2 Suggested answers

1. A: How long has your company been extracting oil and gas from the North Sea?
   B: For 40 years.
   A: How does the crude oil get to the refineries on shore?
   B: Before, it was transported / used to be transported by tankers, but now it's brought ashore by pipeline.
2. A: How long have companies been producing LPG?
   B: Since 1912.
   A: When did motorists start using LPG in their cars?
   B: They started using it in the 1940s as an alternative to petrol and diesel.
3. A: Have sales of LPG in Europe been increasing or decreasing recently?
   B: For the past eight years, sales have been rising / have risen steadily.
   A: In your opinion, what has caused this increase in sales?
   B: These days, more LPG is being used in rural areas for heating and electricity generation.

2 Smart drilling

1
1 organic   2 liquid   3 geological   4 absorbent
5 solid   6 conventional   7 sufficient   8 innovative
9 flammable   10 partial   11 concentrated

2
1. Classified on the basis of their composition, oil shales include carbonate-rich shales.
2. Developed between 1987 and 1991, the most common classification adapts terms from coal terminology.
3. Described as 'terrestrial' (earth) or 'marine' (sea), oil shales are the result of the initial biomass deposit and its environment.
4. Used since prehistoric times, oil shales burn without any processing.
5. Started in France in 1837, industrial mining was followed by further exploitation in Scotland and Germany.
6. Accompanied by an increase in petrol consumption, the mass production of cars helped to expand the European oil-shale industry before 1914.

3 Lasers

1
1. All the topics are mentioned **except**: Discovery of X-rays; Using X-rays to examine metal beams; The use of lasers in medicine.

2
1 lead case   2 cathode   3 tungsten anode
4 electric current   5 electron beam   6 tube   7 heat
8 oil bath   9 X-ray photons   10 X-ray beam   11 filter
12 atoms

3
1. Our aim; to explain what an X-ray is.
2. to talk about; like to take over
3. you have a look at this slide; can see a diagram
4. can now turn to; to ask Magda to take over; going to talk about
5. Ahmed just explained
6. have to say now
7. there are no; hand back

4 Word list

1
1 e   2 h   3 a   4 g   5 i   6 c   7 b   8 f   9 d

2
1 ruby crystal   2 photons   3 atoms   4 absorb   5 emits
6 back and forth   7 partial   8 concentrated

2 Design

1 Spin-offs

1
1 is used	8 was scanned	15 are assembled
2 is presented	9 Initially	16 scan
3 Originally	10 generally	17 permits
4 were used	11 Currently	18 enable
5 repeatedly	12 are issued	19 are used
6 Finally	13 widely	20 normally
7 was taken up	14 assist	

2
1. to check
2. for reordering
3. that/which helps / to help
4. that/which need
5. for ordering / that/which we order
6. that/which are placed
7. to provide
8. that/which is sent
9. that/which identifies
10. to ensure / that/which ensures

2 Specifications

1
1 heavy-duty	6	Comparative / Relative
2 maximum	7	long-lasting
3 relative / comparative	8	sealed
4 enclosed	9	adequate
5 variable	10	adjustable

2
1. … ride, allowing drivers to travel across rough, open ground.
2. … return, without needing to recharge the batteries.
3. … support, enabling drivers to swivel to the side in order to get off.
4. … to use, without requiring physical strength.
5. … ahead, ensuring that the driver is seen from the front.
6. … baskets, providing space for day bags and shopping.
7. … for users when crossing uneven ground.
8. … reflectors, enabling the scooter to be seen from behind.

3 Properties

1   1 b   2 a   3 a

4   1) Get an architect to draw up plans for a modernised building (and brief him/her on the company policy of energy conservation).

2) Fix an appointment with a solar panel installer for a site visit (and get some costs).

2

1	to kick off	7	more ideas
2	Any thoughts	8	That's very
3	let's focus	9	an idea
4	do you think	10	What about putting
5	I suggest that	11	great idea
6	that makes	12	don't we fix

3   Ticks: Solar panel: 1, 3, 8; Roof tile: 1, 4; Membrane: 2, 5, 6, 7; Insulation: 5, 7, 8

4   Solar panel: 13.3%; 20 years
Roof tile: artificial; 100
Membrane: 0.25 mm; along; across; 85 g
Insulation: 45 mm; 25 mm; 1.4 kg; 0.85 kg

4 Word list

1   1 inflatable 2 solar 3 relative 4 sterile 5 adjustable
6 applicable

2   1 compliance 2 detonator 3 Velcro 4 gravity
5 browser 6 panel

3   1st syllable stressed: a<u>de</u>quate, <u>am</u>bient, <u>re</u>lative, <u>ste</u>rile, <u>stun</u>ning, <u>tox</u>ic, <u>va</u>riable
2nd syllable stressed: ad<u>jus</u>table, app<u>lic</u>able, com<u>pa</u>rative, en<u>closed</u>, in<u>fla</u>table
3rd syllable stressed: incre<u>men</u>tal, unres<u>tric</u>ted

Review Unit A

Section 1

1   Suggested answers
1   Where did you do your degree?
(From 1993 to 1996) I did a BSc degree in Physics at Imperial College.
2   What did you study for your MSc?
(From 1996 to 1998) I studied laser applications in materials and medicine.
3   What was the subject of your laser physics PhD?
I did research into crystal geometry and ion salts. (In particular, I studied atomic positions within crystals.)
4   What were you doing at Durban University (after your doctorate)?
(From 2002 to 2004) I was researching the development of an electron gun.
5   What was your job title at the Institute of Laser Information Technologies?
I was Senior Research Scientist (from 2004 to 2010).
6   In which areas of laser applications have you been working?
(From 2004 to now) I've been doing research into laser applications in medicine.
7   What techniques have you developed?
I've developed special tools and control systems for laser applications in knee surgery.
8   How many articles have you written in your career?
I've written (about) 20 articles, mainly on laser applications in medicine.

2   1 The optical beam of a laser when applied to a small spot leads to intense heating.
2   Widely used in manufacturing, lasers enable precise techniques to be used.
3   Compared with mechanical approaches, lasers have many advantages.
4   Used to drill fine, deep holes, lasers bring high processing speeds and reduced manufacturing costs.
5   Medical applications related to outer parts of the body include eye surgery and hair removal.
6   If used for surgery, a laser results in less bleeding and less cutting of tissue.
7   Used for long-distance data transmission, optical fibre communication sends pulses of laser light along fibre optic tubes.
8   Used for range finding and navigation, lasers enable precise position measurements to be made.

3   1 I'd like to start by talking about lasers in manufacturing.
2   If you look at this slide, you will see the escaping photons.
3   I'm going to move on to the next part of my talk.
4   I'll now ask Boris to take over.
5   As you can see in this close-up photo, it's a partial mirror.
6   I think I've covered the main points.
7   Now I'm going to hand back to Mona.
8   Let's move on to the final section of the talk.
9   We've almost run out of time.
10   Would anyone like to ask a question?

Section 2

1   Suggested answers
1   Exterior Insulation and Finishing Systems (EIFS) are multi-layer wall systems that are used on the exteriors of buildings to provide extra insulation.
2   EIFS are excellent systems for insulating solid masonry walls, maintaining the interior dimensions of the building and reducing thermal loss.
3   EIFS have been used for decades to make buildings more energy efficient, reducing heating bills by up to 30%.
4   The bottom layer includes sheets of 4–5 mm foam plastic insulation fixed to a substrate that is attached to the exterior with mechanical fasteners.
5   The middle layer, consisting of a fibreglass mesh covered with a coat of cement-type adhesive, is applied to the face of the insulation.
6   The outer layer, called the 'finish', is a coloured, textured paint-like material that is applied with a flat spreader and which can be rough or smooth in appearance.
7   It is normal for exterior walls to contain moisture, so some EIFS are designed to let water escape, allowing the wall to dry out.

2   Suggested answers

1	stunning	6	easy	11	ambient
2	conventional	7	circular	12	smooth
3	fast-spinning	8	uninterrupted	13	Incremental
4	sterile	9	adjustable		
5	safe	10	variable		

3 Systems

1 Product recall (1)

1 Suggested answer
The extent of the contamination is still being determined, but householders are being advised to boil their water until further notice. It is thought that sewage from the towns of Littlemore and Overmore is still being discharged into the rivers by the overflow system. Farmers with properties adjacent to the River Bourne are being instructed by the Farmers' Cooperative Union not to take water from the river for their farm animals, but to make other arrangements. The public is being advised (by health officials) not to allow their dogs to enter the water, both along the River Bourne and along the Watership Canal. It is reported that all water sports are being suspended (by the County Council), including the Canoe Regatta planned for 19 October.

Extra inspectors are currently being employed by Northern Water to monitor water purity across the county, while water is being extracted (by the company) from the River Bourne to maintain supplies. The incident is being investigated by the National Water Regulator.

2
1 regrets to announce
2 fails to comply with
3 should be resumed
4 As a precaution
5 In the unlikely event that
6 In these unlikely circumstances
7 for any inconvenience
8 Even though
9 under certain conditions
10 will be subject to

2 Product recall (2)

1
1 Murray   2 Barrossa   3 2010
4 faulty **power steering gear** assembly
5 contact local dealer

2
1 the lower pinion bearing
2 on the power steering gear assembly
3 the lower pinion bearing could separate
4 it would permit movement of the pinion shaft
5 11 accidents to date involving injuries; no fatalities so far
6 an intermittent loss of power steering when making left turns, usually at low speeds
7 install a new lower pinion bearing, or replace the whole steering gear assembly.
8 free of charge / no cost

3 Suggested answers
1 The manufacturer stated that there were flaws in the engine valve springs, which could make the vehicle stall.
2 The company issued a press release last week, covering vehicles from the 2008–2010 range.
3 Customers have complained about recent product recalls, which have caused a fall in resale values of used cars.
4 8,500 hybrid cars were recalled last year after tests revealed faulty fuel tanks that caused fuel to spill after rear-end crashes.
5 Customers were being notified of the product recall by emails urging them to contact their local dealer without delay.

6 The air bag deployment signals are not reliable, which could result in the non-deployment of side air bags in frontal collisions.
7 The driver's side air bag inflator could come apart at the weld, preventing full inflation of the air bag.
8 Drivers noticing extra vibration should bring the car to a dealer for service.

3 Controls

1 Suggested answers
1 Both systems control the direction of a vessel's path automatically, without a crew member manually doing the job.
2 The autopilot works electronically, while the windvane operates mechanically.
3 The sensor checks the required compass bearing and instructs changes in direction via the actuator.
4 the deadband
5 the autopilot

2 1 rudder   2 establish   3 relinquish   4 retain
5 override   6 register   7 vane   8 tiller

4 Word list

1 1 gearbox   2 recall   3 fix   4 update   5 rumble   6 input

2 1 comply   2 experienced   3 perceived
4 establishing / determining   5 overrode   6 panicked
7 relinquish   8 kicked in   9 skidded   10 regained

3 actuator, autopilot, campaign, compensation, fix, gearbox, inconvenience, input, intervention, negotiation, patch, pharmacy, precaution, probability, recall, rumble, scare, throttle, update

4 Procedures

1 Shutdown

1
1 F → that accelerates electron beams
2 F → almost up to the speed of light
3 T
4 F → research has been carried out for drugs, diseases and the penetration of metal structures

2 A: linac   D: bending magnets
B: electron gun   E: storage ring
C: booster synchrotron F: beamline

3
1 vacuum
2 a quarter of a millimetre wide; a sixtieth of a millimetre high
3 air molecules
4 one trillion
5 20
6 twice
7 stable; higher average brightness
8 speed of light; powerful magnetic field

2 Overhaul

1 1 Order: D C B A

2
1 explode
2 install
3 emptied
4 examined it
5 do
6 get hotter
7 travelled quickly back
8 removed
9 passed quickly through
10 concentration

3
1. 1 look into
   2 carry out
   3 rule out
   4 Clean out
   5 put off
   6 warm up; carry on; start up
   7 undergo

3 Demonstration

1. 1 metal filler
   2 flux jacket
   3 gaseous cloud
   4 welding tip
   5 filler rod
   6 acetylene
   7 mixing chamber
   8 grounding wire
   9 power source
   10 clamp
   11 electrode lead
   12 electrode
   13 metal filler
   14 weld seam

2. 1 f 2 d 3 b 4 c 5 e 6 a

4 Word list

1. 1 carried out
   2 overhauled
   3 start up
   4 built up
   5 set off
   6 shut down
   7 rule out
   8 undergo
   9 put off

Review Unit B

Section 1

1. 1 registering
   2 panicked
   3 skidded
   4 indicated
   5 establishes / has established
   6 recalled
   7 relinquish
   8 override
   9 interpret
   10 regain
   11 maintain
   12 comply

2. 1 A small above-decks autopilot is inexpensive, but it is not recommended for an ocean-going boat.
   2 Though it is ideal for light conditions, it will have difficulty steering in rough seas.
   3 Above-decks autopilots are easy to fit, whereas below-decks ones need professional installation.
   4 An autopilot will steer a vessel automatically, but it is still necessary to keep a 24-hour lookout on board.
   5 A solo sailor can utilise a windvane in order to sleep. Nevertheless he or she needs to check the wind direction from time to time.
   6 A basic windvane controls the rudder directly, whereas a more advanced system controls the tiller by means of a servo-oar.
   7 Self-steering systems allow a boat to be sailed continuously, although sailing times may be longer than when the boat is sailed manually.
   8 A windvane does not keep the boat sailing on the required compass bearing. Nevertheless, it steers the boat safely at a preset angle to the wind.

Section 2

1. 1 blowdown drum
   2 furnace
   3 splitter tower
   4 outflow valve
   5 level indicator

2. 1 setting off
   2 inlet
   3 came up to
   4 go off
   5 let in
   6 heat up
   7 filled up
   8 turned off
   9 let out
   10 outlet

11 go up
12 overflowed
13 build-up
14 built up
15 brought about
16 started up
17 outburst
18 shut down
19 carried out
20 rule out

5 Processes

1 Causation

1. 1 A supply of iron ore of uniform size and quality is a factor in iron production.
   2 Some difficulties in production are due to the varying iron content and the presence of sulphur and phosphorous in the ore.
   3 In a blast furnace, the introduction of super-heated air gives rise to a reaction where the burning coke raises the temperature to 1,535 °C.
   4 Coke is the fuel which results from heating coking coal without oxygen at temperatures of 1,000–3,000 °C for a period of 18–24 hours.
   5 Carbon monoxide is released from the burning coke, leading to a reaction with the iron ore.
   6 Impurities like carbon and sulphur are removed from the molten iron as a result of the production process.
   7 A steady blast of hot air and gases is forced through the loaded raw materials, resulting in a temperature increase up to the melting point for iron.
   8 Blast furnaces can operate continuously for up to ten years due to their design.
   9 Iron ore, limestone and coke are continuously charged into the top of the blast furnace, resulting in a continuous production process without loss of heat.
   10 The presence of limestone causes the impurities to gather on the surface as slag, where they can easily be removed.

2. 1 emissions
   2 Electrolysis
   3 carbon
   4 Iron oxide
   5 cell
   6 electrolyte
   7 anode
   8 cathode
   9 by-product
   10 oxygen
   11 carbon
   12 sulphur
   13 high-purity

2 Stages (1)

1. 1 contains
   2 is found
   3 differ
   4 makes
   5 are created
   6 is used
   7 is produced
   8 is cooled
   9 burns
   10 support

2. 1 August 2004
   2 May 30th, 2007
   3 Pohang, South Korea
   4 1.5 million tonnes per annum (mta)
   5 Orissa State, India
   6 4 mta
   7 2010
   8 12 mta

	Traditional iron making	FINEX Process Technology
1	Yes	No
2	Yes	No
3	iron ore pellets Cost: 100%	powdered iron ore Cost: 23% cheaper
4	high-class soft coal Cost: 100%	general coal Cost: 20% cheaper
5	N/A (not applicable)	Indian iron ore with high aluminium content
6	100%	10–15% lower
7	More	Less
8	100%	fewer maintenance staff; less production time

3 Stages (2)

1
1 hopper
2 smelting pot
3 carbon blocks
4 molten electrolyte
5 iron bar
6 cathode
7 anode
8 tap hole
9 ladle

2 Suggested answers
1 What is aluminium made from?
2 The main piece of equipment is the smelting pot, which is made with an outer shell of steel.
3 How much electricity does a smelting plant use?
4 How hot does the electrolyte (in the smelting pot) get?
5 It ends up at the bottom of the pot.
6 How is the molten metal extracted from the pot?
7 Can the smelting process be halted?
8 How often do the carbon blocks in the anode have to be replaced?
9 Which gases are produced during the smelting process?
10 What are the properties of aluminium?

4 Word list

1
1
1 Inert gas pipes are used in a BOS converter to agitate the mixture.
2 An oxygen lance is used to heat the mixture and melt the scrap.
3 Lime is added to the converter to bond with impurities.
4 Residual slag is removed from the converter and is recycled.
5 A sub-lance is lowered into the converter to measure the carbon content of the molten metal.

2
1 The mineral bauxite is extracted from the ground.
2 Rocks of bauxite are pulverised in a crusher.
3 Ground bauxite is dissolved in sodium chloride in a digester tank.
4 Pure alumina crystals sink to the bottom of the precipitator tanks.
5 The crystals are calcinated in rotary furnaces.

2 1st syllable stressed: molecule, oxygen, shortage, coolant
2nd syllable stressed: converter, electrolyte, filtration, hydroxide, precipitator
3rd syllable stressed: aluminium, electrolysis, supervision
4th syllable stressed: precipitation

6 Planning

1 Risk

1
1 are likely to be
2 virtually certain
3 The probability that
4 is no doubt that
5 could happen
6 There's a strong probability
7 is almost certain to be
8 likely to happen
9 a high possibility
10 strong likelihood that
11 would be able to
12 no doubt that
13 now a slim chance that

2
1 minimal
2 imminent
3 impressive
4 disastrous
5 potential
6 insignificant
7 critical
8 minor

2 Crisis

1
1 oil slick
2 containment
3 skim
4 leak
5 surface
6 methanol
7 relief well
8 boom
9 main riser
10 fracture
11 pump
12 subsea cap
13 well head
Vertical word: insertion tube

2
1 A new task force is going to be set up to deal with future subsea oil spills in the Gulf.
2 Underwater well containment equipment will be made available to all oil and gas companies operating in the Gulf.
3 A new rapid-response oil spill containment system is going to be developed to help prevent another disaster like the 2010 blowout.
4 The rapid-response system will be made available for mobilisation within 24 hours.
5 It will be used on a range of equipment and in a variety of weather conditions.
6 By the end of this year, the new operations centre will have been set up in a permanent facility in Houston, Texas.
7 A spokesperson says that contracts for capture vessels are on the point of being signed by officials.
8 In addition, a full range of manifolds, jumpers and risers are about to be ordered and delivered to the operations centre.
9 In the meantime, existing equipment will be assessed for use in the short term.
10 Technical personnel with experience from the 2010 oil spill are going to be retained by the new rapid-response project team.

3 Project

1
1 70,000
2 Yes
3 15
4 Europe, Russia, USA
5 1) building new plants 2) upgrading existing plants 3) extending their working life

2
1 Construction cost: 125 billion rand = **$17.5** bn; Power output: **4,800** MW
2 Number of reactors: **5–6**
3 Need more **experts** to be trained; Need for support industry for **construction**, **maintenance** and **waste**
4 One NPP costs **3 times** more than a coal-fired plant; Cheaper to build **a series of** NPPs than a single one

5 Locally produced **uranium** already available; Plan to **convert uranium** in RSA for NPPs
6 Generation of an extra **1,025** MW
7 **Higher** construction costs, **higher** operating and maintenance costs, **lower** fuel costs

3 1 (1) aesthetically stunning (2) economic powerhouse (3) an important manufacturing hub (4) land reclamation programme (5) artificial island; reactor installation
2 (1) potential radiation leaks
(2) seawater contamination
3 (1) global sea levels (2) at risk of flooding

4 Word list

1
1	strategy	6	consequence
2	consideration	7	likelihood
3	magnitude	8	certainty
4	agenda	9	probability
5	concern	10	doubt

2
1	well head	6	insertion tube
2	subsea cap	7	relief well
3	oil slick	8	marine biologists
4	underwater robot	9	risk assessments
5	remotely-operated underwater vehicle / ROV		

Review Unit C

Section 1

1
1	as a direct result of	6	due to
2	caused	7	direct cause of
3	As a result of	8	gave rise to
4	cause	9	resulting in
5	causing it to	10	leading

2 Suggested answer
Every 20 days, the anodes have to be replaced, as the two carbon blocks forming the anode have shrunk to one third of their original size. An operator uses a gantry crane to lift the anode assembly from the top of the pot and place it on a rack for cooling. The two three-metre rods that hold the carbon blocks are later re-used. While the top of the pot is open, waste matter from the crust is removed with a pincer crane and dumped in a truck. Next, the pot is recharged from above with a fresh supply of alumina crystals, Finally, a new anode assembly is placed on top of the pot, so that the smelting process can continue without a break.

Once a day, molten aluminium is siphoned from the potlines by the operators. A collector tank is moved into place to siphon off the molten metal. A thick tube is inserted through a hole in the side of the pot, and air is extracted from the collector tank. This creates a vacuum, allowing the molten metal to be siphoned off.
In the casting house, some of the molten aluminium is poured directly into moulds, where it slowly cools and hardens. The rest of the aluminium passes into furnaces where it is mixed with other metals to form alloys. Gases such as nitrogen and argon are used to separate impurities and bring them to the surface where they can be skimmed off. The resulting purified alloys are then poured into moulds and cooled rapidly. Water sprays speed up the cooling process.

Section 2

1
1	serious	6	breathtaking / significant
2	upbeat	7	critical / significant
3	significant / breathtaking	8	imminent
4	potential	9	catastrophic
5	minimal		

2 (a) slight (b) strong (c) highly (d) virtually

3
1 are going to be defined
2 will be prepared / will have been prepared
3 are going to be claimed
4 is on the point of being drawn up and discussed
5 are about to be filed
6 will have been collected and presented
7 will be divided up
8 is going to be carried out
9 will be speeded up
10 will have been made

7 Developments

1 Progress

1 1 g 2 c 3 e 4 a 5 h 6 b 7 d 8 f

2 Model answer
Aim: New cordless handheld computers are being developed to overcome the limitations of the smartphone screen size.
Components: The system consists of a large screen (25 mm diagonal), a built-in lithium-polymer battery with up to 10 hours of use-time, and a solid-state flash processor with a choice of 16 GB, 32 GB or 64 GB capacity.
Operation: Features are available once inside a particular app. Users can 'pinch, swipe and tap' the interactive screen while navigating an app. There is a 12 mm-wide border around the active display screen. This is useful for holding the Voyager while keeping fingers off the active display.
Method: The device uses a 1 GHz ARM processor, which includes integrated 3D graphics, audio, power management, and storage to load apps and web pages from the internet.
Outdated technology: In the past, a stylus was used on an interactive screen with handwriting recognition software. In addition, smartphones had the disadvantage of a very small screen.
Recent development: A virtual keyboard is provided for the Voyager. As a result, tap-typing lots of text is much easier than on a smartphone.
Work in progress: At the moment, the home screen interface is being redesigned so that it can incorporate user-controlled options. Also, a GPS live-data program is being adapted so that it can be installed as a standard app on the Voyager.
Further work required: The Voyager needs to be fitted with a camera so that it can be used for video conferencing as well as online chatting. This work is planned for the next model.
Future target: More people are going to turn to the cordless handheld format for their portable computing needs. As a result, the number of available apps will no doubt increase rapidly.

2 Comparison

1
1	resistive	6	clarity	11	durability
2	electrical	7	resistive	12	resistive
3	capacitive	8	capacitive	13	durable
4	conductivity	9	up-to-date	14	light
5	metallic	10	sharp		

2
1 Unlike other manufacturers, Misawa delivers top-class picture quality across the price range.
2 Instead of moving away from plasma, Misawa believes that this format is best suited to 3D TV.
3 While picture clarity may not improve this year, its new TVs will have greater energy efficiency.
4 The new Sakatas are much easier to use than the outdated ones.
5 The screen of the Sakata 201 is slightly more reflective than that of the Sakata 101.
6 Expensive models have good acoustic quality compared with the poor acoustic quality on cheaper models.
7 You need a higher frame rate when watching sport, although the software can make the picture look unnatural.

3 Product

1
1, 2 T
3 F → Lasers are more powerful but more variable light sources than LEDs.
4 T
5 F → Copper wires need more relay systems than fibre optics.

2
1 or; that is
2 To put this in everyday language
3 to put that another way
4 Another way of putting it is
5 that is to say
6 that is

3
1 greatly
2 be greater than
3 get rid of
4 light flashes
5 less pure
6 impossible to decode
7 shown
8 restored
9 long-distance

4 Word list

1
1 stylus
2 wireless broadband
3 static data
4 contaminant
5 contact point
6 gesture
7 durability
8 projector
9 live data
10 conductivity

2
1 data, capability, limitation, radiation, label
2 reality, clarity, contaminant, capacitive, metallic
3 digital, disaster, surface

8 Incidents

1 Missing

1
1 f 2 b 3 g 4 d 5 a 6 e 7 c

2
1 thefts
2 incident
3 pallets
4 inventories
5 outcome
6 access
7 (had) disabled
8 rule out
9 insider
10 assumption
11 stressed
12 tightened up
13 speculated
14 threat

3 Suggested answers
1 The theft could have been carried out by an insider.
2 A single person can't have carried out the theft unaided.
3 The drugs couldn't have been removed during the normal working day.
4 The movement sensor must have been disabled at the beginning.
5 We ought not to have stored the pallets of drugs in the low-security zone.
6 Overhead security fittings should have been routinely inspected.

7 CCTV and security lights ought to be installed for night-time use.
8 All the security arrangements should be reviewed and tightened up.

2 Confidential

1
1 F→ …, it may cause harm by changing control programs.
2 F → The Stuxnet worm is complex and larger than average.
3 T
4 F → … and can carry out unauthorised actions within a network.
5 F → … for the intentional and harmful distribution …

2
1 copying
2 encoded
3 noticed
4 activates
5 immediate
6 delete
7 company
8 short
9 private
10 permitting
11 carrying out
12 decoded

3
1 I would like to know how many attempts there have been to interfere with our company computer system.
2 Please check if/whether anyone has used the computer who has not been individually authorised.
3 It's important that we determine if/whether any files have been accessed or modified.
4 I need to find out who has been trying to obtain usernames or passwords of other users.
5 We've got to discover if/whether anyone has copied software without permission or not.
6 Tell me how much money has been charged for private use of computer services.
7 It's vital that we determine if/whether anyone has knowingly introduced a worm or harmful program into the computing facility.
8 Do we know if/whether anyone used the computing facility for a criminal act?
9 We've got to find out if/whether anyone is sending bulk emails from our computers.
10 Please confirm how many spot checks we made last month and what we discovered.

3 Danger

1
1 emit, emission
2 corrode, corrosion/corrosivity, corrosive
3 flammability, flammable
4 hazard, hazardous
5 infect, infection, infectious
6 radioactivity, radioactive
7 react, reaction, reactive
8 toxicity, toxic

2
1 Lowermoor Water Treatment Works
2 unstaffed; locked
3 aluminium sulphate
4 to collect impurities (as part of water treatment process)
5 drinking water
6 30,000
7 620,000 micrograms per litre
8 200 micrograms per litre

3,4
1 ✓ Without a doubt / 60,000 fish killed; soft mud in pipes stirred up
2 ✗ On the contrary / water authority insisted water was safe to drink; suggested mixing the water with orange juice
3 ✗ Definitely not / acid corroded the network and contaminated the water supply with high levels of copper, zinc and lead
4 ✗ Actually / boiling the water made it more toxic; it concentrated the contaminants and heavy metals

5 ✓ In fact / water tank should have been cleaned out every six months; hadn't been cleaned for 3 years, and this resulted in build-up of mud, which went into the water supply network

6 ✓ To some extent / routine cleaning of tanks not done; inadequate health and safety systems; tanks not clearly labelled

4 Word list

1
1	out-of-contact	5	wake-up-and-wipe
2	event-based	6	pre-set
3	time-based	7	pre-scheduled
4	on-demand		

2 1st syllable stressed: <u>in</u>cident, <u>in</u>ventory, <u>cor</u>porate
2nd syllable stressed: as<u>sump</u>tion, in<u>sid</u>er, sor<u>ta</u>tion, tox<u>ic</u>ity, en<u>cryp</u>ted, pre<u>lim</u>inary, un<u>au</u>thorised
3rd syllable stressed: instan<u>ta</u>neous
4th syllable stressed: consoli<u>da</u>tion

Review Unit D

Section 1

1 Model answer

1 The purpose of this brief report is to compare two 3D TV systems – the Active Shutter 3D Display and the Passive Shutter 3D Display – in order to assist the company in the process of selecting the best 3D TV system for our hotels.

2 Both types of 3D TV send a different picture to each eye. The brain then combines them into one 3D image through the use of either active or passive shutter displays. Both systems have very high clarity because they use top-of-the-range TVs. This means that there are some cost disadvantages for both TV systems, however. Both systems show two images on the screen, and both require the viewer to wear glasses. In the case of one system, the glasses need a power source, while in the other system, they don't.

2 Model answer

3 With the Active Shutter (3D display) system, the TV shows the same two images, but not at the same time. The TV flicks between showing a full-screen image for the left eye and a full-screen image for the right eye. The glasses open and close shutters at the same speed to ensure that each eye sees the correct picture.
With the Passive Shutter (3D display) system, two images are shown on the TV at the same time. Each lens of the polarised 3D glasses allows only one of the polarised images through, filtering out the other.

4 These differences in the technology give rise to differences in performance between the two systems. In the Active Shutter system, the glasses blink on and off. Since the glasses need a power source to operate, this makes them heavy. In addition, the glasses must be suitable for their own brand of TV.
In the Passive Shutter system, glasses are lighter and cheaper. However, as each eye only sees half the pixels on the screen, this means that the 3D content won't be shown in full HD.

3
1	Since	4	In addition
2	On the other hand	5	results in
3	Whereas	6	For these reasons

Section 2

1
1 I can't understand what can have happened. …
2 Might someone have used your password?
3 No, they couldn't have done that, …

4 Could they have interfered with an ATM in the high street?
5 Somebody might have installed a camera …
6 No, that can't have happened.
7 I think I know what must have happened.
8 I shouldn't have replied, but I did.
9 It must have been a fake email.
10 You ought not to have replied, and you should have ignored the request.
11 Ought I to tell the bank?
12 Yes, you should go to your bank …
13 The money might be retrieved, if you're quick.
14 But you may be unlucky and you may have lost the lot.

2
1 Why didn't you check the email properly?
2 Was that email from the tax office or not?
3 When did the money go out of the account?
4 Has anyone ever written down their PIN or password?
5 Has she or has she not been destroying all bills and receipts using the office shredder?
6 Have you ever disclosed personal or security details in reply to an email?
7 Was all the money transferred from our Savings Account to our Current Account?
8 Was he using the same password for more than one account?
9 Has anybody replied to an urgent email asking them to confirm their bank details?
10 When did we last have our anti-virus software updated to protect our computer network security?

9 Agreements

1 Proposals

1 1f 2h 3g 4e 5a 6d 7c 8b

2
1 both
2 by security patrols on foot
3 he is concerned about the installation costs, but decides to go for a total security package
4 both
5 not yet (but it will be submitted following the meeting)

3
1 keep an eye on all vehicle movements
2 incoming vehicles make their way
3 cut down on the salary bill
4 should we go about installing
5 be able to home in on areas
6 go for a total security package
7 running down its operations
8 could you draw up a detailed proposal

2 Definitions

1 Suggested answers
1 A thermostat is a device/sensor for measuring variations in temperature.
2 GNSS is a system that provides global geo-spatial positions.
3 A tyre pressure gauge is a gauge / an instrument for measuring air pressure in vehicle tyres.
4 A CO alarm is a detector/sensor that measures the concentration of CO.
5 The ISO is an organisation for setting international standards.
6 Electroplating is a process that coats metal objects with another metal.
7 A moisture meter is an instrument which measures humidity in substances.
8 A carbon sensor is a sensor which measures the carbon content of molten metal.

2
1 A moisture meter is a portable battery-operated instrument with two electrodes which measures humidity in substances by measuring the electrical resistance between two points.
2 A tyre pressure gauge is a gauge for measuring air pressure in vehicle tyres, consisting of a flattened thin-wall closed-end tube, connected at the open end to a pipe containing the air pressure to be measured.
3 The ISO is a voluntary organisation for setting international standards for trading categories of goods in a way that allows and encourages fair competition.
4 GNSS is a navigational system that uses GPS co-ordinates to provide global geo-spatial positions by means of electronic receivers.
5 A carbon sensor is a device mounted on the end of a lance which measures the carbon content of molten metal in a converter during the steelmaking process.
6 A CO alarm is a small battery-operated detector containing an electrochemical fuel cell that measures the concentration of CO (in the atmosphere) and emits an audible alarm when the concentration exceeds a safe limit.
7 Electroplating is a plating process which uses an electrical current to coat metal objects with another metal by immersing them in a bath of electrolyte.
8 A thermostat is a heat-sensitive electrical device, consisting of two dissimilar wires, for measuring variations in temperature.

3 **Across:** 8 photoelectric 3 spatial 7 dissimilar
10 structural 1 infrared 6 metallic 4 reliable
Down: 5 precise 2 optional 9 complex

3 Contracts

1
1 F → ... there are a few changes.
2, 3 T
4 F → No, but if there is a disagreement between the contractor and client about quality standards, an independent surveyor can assess the work.
5 F → ... the contractor will require payment for pumping out the water.

2
1 otherwise the contract will be cancelled
2 In the event of bad weather
3 Even if the delivery of building materials
4 on condition that we are satisfied
5 with the proviso that
6 as long as notice is given
7 In case of disagreement over
8 provided that the surveyors

3 1 d 2 i 3 c 4 a 5 g 6 h 7 b 8 e 9 j 10 f

4 Word list

1 1 c 2 a 3 c 4 b 5 a 6 c 7 a 8 b

2 de**fault**, eco**nom**ic, ge**ner**ic, in**tru**sion
cancel**la**tion, **cir**cumstance, com**mence**ment, in**sur**ance, pro**vi**so
dis**sim**ilar, **op**tional, pros**pec**tive, pyro**elec**tric, re**li**able

10 Testing

1 Test plans

1
1 massive
2 reinforce
3 seismic
4 displacement
7 damper
8 significant
9 dissipate
10 magnitude

5 simulation
6 embed
11 anchor rod
12 cushion
Vertical word: acceleration

2
1 three-test
2 15-storey
3 50-tonne
4 carefully-controlled
5 Easy-to-build
6 wood-frame
7 earthquake-resistant
8 fast-growing
9 million-dollar
10 highly-respected
11 labour-saving
12 cost-cutting

2 Test reports

1 Suggested answers
1 A new type of load-bearing wall is needed which will not collapse during severe earthquakes.
2 The experiment was designed to test a sideways force up to magnitude 7.3.
3 One wall sample was reinforced, but the other was unreinforced. Wall displacement was measured by motion sensors attached to the walls and the base.
4 Push-over tests were set up, and the two walls were tested together.
5 The greatest stress was expected on the fifth floor of the building. There was no visual sign of damage.
6 The unreinforced wall is expected to resist 7.3 magnitude earthquakes.
7 The reinforced wall will be subjected to further tests up to 8.0 magnitude.
8 The earthquake simulation lasted 7.58 seconds at a magnitude of 7.3.

2 A Test objective   C Procedure
B Results   D Experimental setup

3 E Conclusions   F Introduction
E metal fatigue, weld seams, explosions, corroded, replacement
F vessel, catastrophic, building up, four

3 Test methods

1–2 Bridge
1st choice: ultrasonic; Reasons: penetrates deep into the girder; no access problems; cheaper
2nd choice: radiographic (X-ray); Reasons: problem of access to top and bottom surfaces of some girders; more expensive
Unsuitable: eddy current; Reasons: flaws may be too deep to detect
Unsuitable: magnetic particle; Reasons: flaws may be too deep; doesn't work well with painted surfaces
High-rise apartment block
1st choice: eddy current; Reasons: flaws close to surface and at right angles to direction of coil winding
2nd choice: radiographic (X-ray); Reasons: flaws may be too thin, may run parallel to X-rays and may not show up on film
Unsuitable: ultrasonic; Reasons: flaw may be parallel to sound waves, may not cause enough reflection
Unsuitable: magnetic particle; Reasons: visual inspection may not be possible

3
1 start the meeting now; first item on our agenda
2 give us some input; find out about this subject
3 extremely useful input; add anything to that
4 briefly summarise what we've discussed
5 to take a decision
6 move on to the next item
7 a very interesting presentation; comment at this point
8 throw this item open
9 decide on some action
10 concludes our meeting

4 Word list

1

1	experimental	7	destructive
2	shake table	8	non-destructive
3	displacement	9	significant
4	acceleration	10	seismic
5	dampers	11	simulated
6	massive	12	long-running

Review Unit E

Section 1

1
1. on condition that
2. in compliance with
3. satisfactorily
4. except when something else is agreed; before
5. in the case of
6. If a disagreement occurs
7. except
8. as long as

2

1	make our way towards	5	home in on
2	went for	6	following up on
3	keep an eye on	7	putting forward
4	cut down on	8	go about

Section 2

1 (ticks for)
1. magnetic particle
2. eddy current
3. ultrasonic, radiographic
4. radiographic
5. ultrasonic
6. ultrasonic, radiographic
7. eddy current, magnetic particle
8. radiographic
9. magnetic particle
10. ultrasonic
11. magnetic particle

2 Suggested answers
Lines 1–3: Ultrasonic and radiographic testing methods can be used with all types of materials, whereas magnetic particle testing can only be used with ferromagnetic materials, and eddy current testing can only be used with conductive materials.
Lines 4–5: Radiographic testing can detect, locate and measure flaws, while ultrasonic testing can only detect and locate flaws.
Lines 6–7: Ultrasonic and radiographic testing can detect flaws at various depths, but eddy current and magnetic particle testing can only detect flaws on or near the surface.
Lines 8–9: In magnetic particle testing, the operator must have access above the flaw. However, in radiographic testing, the operator must have access above and below the flaw.
Lines 10–11: In ultrasonic testing, the test instrument must contact one surface of the material, but in magnetic particle testing the test instrument must contact the surface of the material above the flaw.

3 1 D 2 C 3 B 4 A

4 Suggested answers
1. The use of three-dimensional modelling software today allows designers to create extraordinarily complex forms. An external, low-energy LED source throws light onto a leaf-shaped element made of thermo-formed PMMA.

2. The chair is made entirely of polypropylene, strengthened with glass fibre. Nitrogen is injected at high pressure during the moulding process, reducing the length of the production cycle and introducing an internal air cavity requiring less material. The three-minute cycle time demonstrates the impressive efficiency of gas-assisted injection moulding.
3. This range of items is produced by a mobile machine that extrudes spaghetti-like strands of plastic. The bowl exhibited here was made of a recyclable polymeric resin developed specifically for profile extrusion requiring high clarity and good resilience.
4. The planters were designed by solving their complex geometry with three-dimensional digital modelling. The lacquer-coated objects are made of rotational-moulded polyethylene similar to that used in producing large rubbish containers.

11 Accidents

1 Investigation

1
1. Kolontar, Hungary (about 160 km from capital Budapest)
2. 4th October, 2010
3. 9 dead, 120 injured, 700 evacuated
4. dam wall not strong enough to deal with increased capacity caused by heavy rain; unsuitable site for reservoir (because of unstable soil)

2
1. 7; 40; rivers; farmland
2. 1,100
3. 10 hectares; 20–30 million
4. 600,000–700,000
5. protective wall: 620 m; 3 m
6. toxic chemical; dries out
7. construction
8. recycled; storage reservoir
9. 30 years
10. unstable soil

3
1. The reservoir should not have been constructed with an earth dam on an unsuitable site.
2. The possibility of heavy rains should have been taken into account when the capacity was determined.
3. The structure of the dam wall should have been inspected periodically.
4. The red mud should have been recycled after drying out.
5. The company should be taken into public ownership for a period of two years.
6. All documents concerning the reservoir should be inspected by the disaster agency, in case negligence was involved.
7. Protective masks and outerwear should be worn by workers until further notice.
8. The concentration of alumina dust in the air should continue to be monitored by the disaster agency.

2 Report

1
Group 1: 1 d 2 h 3 f 4 a 5 c 6 g 7 e 8 b
Group 2: 9 k 10 m 11 j 12 o 13 l 14 i 15 p 16 n

2
1. The crash would not have happened if Gran Canaria Airport had not been closed.
2. If the controller had been able to see the aircraft on the ground, he could have given clearer instructions.
3. If the airport had been equipped with ground radar, the controller would have been able to monitor the aircrafts' positions in poor visibility.

4 If communication between the controller and the pilots had been better, the KLM captain would not have thought that he had clearance for take-off.

5 If the KLM flight engineer had repeated his concern about the controller's instruction, the pilot might have abandoned his take-off.

6 The pilots would have been able to see each other if the fog had been less thick.

7 The fire service would have gone immediately to both planes if they had been able to see them.

8 If the KLM plane hadn't taken on extra fuel, the blaze might have been less severe.

3 Communication

1 1 f 2 d 3 g 4 c 5 e 6 a 7 b

2 1 Excuse me   4 flood alert   7 worsening
   2 emergency   5 full
   3 definitely   6 duties

4 Word list

1 1 Extensive   5 debris   9 conclusions
   2 dam   6 contaminated 10 Preliminary
   3 reservoir   7 toxic   11 abstract
   4 plant   8 findings   12 attachment

2 1 backup, dam, attachment, abstract, analyse
   2 filtration, vibrate, indicator, station, intake
   3 hall, fuselage, fatigue

12 Evaluation

1 Projects

1 1 evaluation   9 recuperator
   2 aim   10 turbine
   3 electricity   11 generator
   4 air   12 peak-period electricity
   5 off-peak electricity   13 requirements
   6 grid   14 generate
   7 compressor   15 rely
   8 limestone cavern   16 criteria

2 Performance

1 1 F → The results are complete and were ready last week.
   2 T
   3 F → It was possible to modify the test schedule …
   4 F → The researcher has learnt three lessons …

2 1 written (but late)
      2.1 slow in booking up
      2.2 measuring instrument
      2.3 underestimated time (needed); test equipment
      2.4 one extra (unnecessary)
   3 new range of batteries has increased efficiency
   4 two tests which didn't need the missing (measuring) instrument
      5.1 test specification (before starting each test)
      5.2 contingency plan (in case rescheduling of tests is necessary)
      6.1 laboratory schedules (at outset of project)
      6.2 list of equipment needed (as early as possible)
      6.3 any extra (items of) equipment

3 1 hanging on; got on with
   2 turn out
   3 come up with
   4 go about
   5 get hold of; put off; set up
   6 came up
   7 went ahead with

3 Innovation

1 1 nanoscale   7 nanoparticles
   2 nanometres   8 graphene
   3 smart dust   9 graphene sheet
   4 Nanowires   10 graphite
   5 nanogenerator   11 nanoscale pores
   6 nanofluid

2 **Diagram:**
   1 less than 1 mg   3 1 m^2
   2 4 kg   4 0.77 mg
   Table:
   2 hexagonal / six-sided structure
   3 thin, strong, flexible, stretchable, conductor; thermal conductivity: high; conductivity: high
   4 70 cm
   5 touch; light; solar
   6 flexible; gas; satellites; aircraft
   7 42 N/m; 100 times
   8 (see diagram on the left)
   9 non-coloured; 2.3%; almost
   10 copper
   11 10 times better than copper

4 Word list

1 1 criteria   6 appraisal/assessment
   2 performance   7 achievement
   3 assessment/appraisal   8 effort
   4 participation   9 evaluation
   5 assertiveness/achievement   10 scope

2 achievement, appraisal, co-operation, contingency plan, criteria, evaluation, gravity, nutrient;
   objective, participation, performance, schedule, water vapour;
   nanogenerator, transparency

Review Unit F

Section 1

1 1 C 2 A 3 F 4 B 5 H 6 I 7 D 8 G 9 E

2 1 If the operating company had had good organisation and management, the aircraft would have been maintained in good operating order.
   2 The airline would not have been allowed to operate unsafe aircraft if the regulatory authority had taken its safety responsibilities more seriously.
   3 If the manufacturer had responded to previous pressurisation incidents in the same type of aircraft, the malfunction would not have occurred.
   4 The plane would have crashed much sooner if it had not been in autopilot mode.
   5 If the crew had shared a common language, they would have been able to communicate better with each other and with the maintenance department.
   6 If they hadn't found the FDR, the causes of the accident would have been less clear.

Section 2

1 a 5 b 3 c 6 d 1 e 4 f 2

2 1 there was no need   9 drops
   2 needed to   10 nutrients
   3 were not obliged   11 objective
   4 essential   12 growing
   5 gravity   13 had to be developed and installed
   6 having evaporated
   7 water vapour   14 is turning out / has turned out
   8 produce   15 greenhouses